Independent TExES Science 7–12 Exam Study Guide 2027–2028:

700 Questions with Detailed Answers for Chemistry, Biology, Physics, and Earth Science Domains

Caldwell Abiodun

Author Introduction

Preparing to teach science in Texas demands more than subject knowledge—it requires mastery of pedagogy, state standards, and the ability to translate complex scientific principles into meaningful classroom learning. The author of this TExES Science study guide approaches that challenge with a clear mission: to equip aspiring educators with the confidence, competence, and strategic insight necessary to succeed on the TExES Science certification exam and thrive in Texas classrooms.

With extensive experience in science education and educator preparation, the author brings a deep understanding of both the content domains assessed on the exam and the instructional frameworks that guide effective teaching. This expertise is strengthened through close collaboration with a dedicated team of science educators, curriculum specialists, and assessment experts who collectively analyze exam objectives, competency standards, and emerging trends in science instruction. Together, they ensure that every chapter reflects accuracy, alignment with Texas Essential Knowledge and Skills (TEKS), and practical classroom relevance.

The TExES Science exam evaluates not only foundational knowledge in life science, physical science, earth and space science, and scientific inquiry, but also the ability to apply that knowledge through sound instructional decision-making. Recognizing this dual focus, the author has crafted this guide to bridge theory and practice. Readers will find clear explanations of core scientific concepts, structured review of tested competencies, and carefully designed practice questions that mirror the rigor and format of the actual assessment. Emphasis is placed on critical thinking, data analysis, and real-world application—skills essential for both exam success and effective teaching.

Understanding the pressures candidates face, the author prioritizes clarity, organization, and strategic guidance. Complex topics are broken down into manageable components, common misconceptions are addressed directly, and test-taking strategies are integrated throughout. The goal is not simply to help candidates pass an exam, but to strengthen their professional foundation as science educators committed to inquiry-based learning and student engagement.

This study guide represents a shared commitment to academic excellence and educator readiness. Through thoughtful preparation, deliberate practice, and a strong grasp of scientific and pedagogical principles, candidates can approach the TExES Science exam with assurance—prepared not only to earn certification, but to inspire the next generation of scientific thinkers in Texas classrooms.

Table of Contents

Chapter 1 — Scientific Inquiry and Processes

The Nature and Purpose of Scientific Inquiry

Scientific inquiry is the deliberate process through which phenomena are observed, interpreted, and explained according to systematic reasoning. It represents the disciplined pursuit of knowledge grounded in empirical evidence, allowing science to distinguish itself from speculation or belief. Inquiry involves both the cognitive act of questioning and the methodological framework of testing those questions through observation, experimentation, and analysis. Within the educational and professional practice of science, inquiry operates as both a method and a mindset. It demands accuracy, intellectual humility, and the willingness to revise conclusions when evidence warrants. The purpose of inquiry is not simply to confirm prior assumptions but to refine understanding through falsifiable testing. By emphasizing cause and effect, reproducibility, and objectivity, inquiry ensures that knowledge claims remain verifiable. In the TExES framework, this concept defines the foundation of scientific literacy. A teacher's understanding of inquiry includes the ability to model how hypotheses evolve, how data inform conclusions, and how uncertainty contributes to progress. Through this process, inquiry becomes both an epistemological principle and a practical guide for developing scientific competence in learners.

Historical Evolution of Scientific Methodology

The modern scientific method is the result of centuries of intellectual development, evolving from philosophical reflection to formalized experimentation. Early inquiry traditions can be traced to Greek natural philosophy, where observation was first recognized as essential to understanding nature. The Renaissance revived empirical reasoning, giving rise to controlled experimentation through figures such as Galileo and Bacon, who established induction and systematic observation as cornerstones of reliable knowledge. The Enlightenment extended these principles through quantification and replicability, culminating in the experimental frameworks that define modern science. In the nineteenth and twentieth centuries, developments in chemistry, biology, and physics introduced specialized methods—statistical analysis, hypothesis testing, and modeling—that diversified the scientific approach. By the late twentieth century, inquiry had expanded to include computational simulation and interdisciplinary investigation. Within education, this historical progression informs the way students learn to question and test their surroundings. Understanding the lineage of methodology enables teachers to frame scientific reasoning as an evolving dialogue rather than a fixed formula. In the classroom, historical awareness underscores that the scientific method is not a rigid sequence but a flexible framework adaptable to new evidence and contexts. This awareness cultivates respect for both the tradition and creativity inherent in scientific discovery.

Phases and Logic of Scientific Investigation

Every investigation begins with a question rooted in observation. From that question, scientists generate hypotheses—proposed explanations subject to empirical testing. Data collection follows, shaped by careful planning to minimize bias and ensure reliability. Once results are obtained, analysis determines whether the evidence supports or refutes the hypothesis. Interpretation connects findings to existing theories, and communication allows the wider community to evaluate, replicate, or challenge conclusions. The logic underlying this sequence is cyclical rather than linear. Each conclusion generates new questions, propelling inquiry forward. Central to this logic is falsifiability: a hypothesis must be structured so that it can, in principle, be proven wrong. The distinction between correlation and causation, the control of variables, and the use of appropriate instrumentation all reinforce the integrity of this process. In the classroom, demonstrating these phases through laboratory work or data-based activities mirrors authentic scientific reasoning. By emphasizing precision in measurement and clarity in argumentation, teachers help students internalize the logic that makes inquiry both credible and self-correcting. This systematic cycle reflects the essential reasoning tested within the TExES Science 7–12 framework.

Empirical Observation and Measurement

Observation forms the empirical foundation of all science. It involves disciplined attention to phenomena and the capacity to record details without imposing premature interpretation. Measurement extends observation by translating sensory data into quantitative terms, allowing comparison, prediction, and statistical evaluation. Instruments—ranging from simple rulers and thermometers to advanced spectrometers and telescopes—serve as extensions of human perception. Accuracy, precision, and calibration are therefore not procedural details but ethical responsibilities in maintaining the credibility of evidence. In the educational setting, the teaching of measurement reinforces objectivity and consistency, bridging the conceptual with the practical. Students learn that data reliability depends on standardization and repeatability, not on the persuasive strength of personal conviction. Through repeated measurement, error analysis, and graphical representation, learners develop an intuitive understanding of uncertainty and variability. Empirical practice thus fosters both technical skill and intellectual discipline. The TExES assessment reflects this through questions that require the interpretation of numerical data, analysis of error margins, and selection of appropriate tools for specific investigative purposes. The act of observing and measuring accurately embodies the essence of scientific professionalism.

The Role of Hypothesis and Theory Formation

Hypotheses serve as the pivotal link between observation and explanation. A hypothesis is not a guess but a reasoned proposition grounded in existing knowledge, formulated in such a way that it can be empirically evaluated. When repeatedly supported by evidence and consistent with related findings, a hypothesis may contribute to the development of a theory—a coherent system of ideas that explains a range of phenomena. Theories differ from laws, which describe patterns mathematically without necessarily explaining their causes. In the context of teaching and testing, understanding this distinction is essential to avoid misconceptions such as viewing theories as "unproven facts." The scientific enterprise advances by continuously testing and refining theories in light of new evidence. This dynamic process ensures that science remains both stable and adaptable. For exam preparation, candidates should recognize how theoretical frameworks guide investigation design and interpretation. Theories unify data across disciplines and help predict outcomes under varying conditions. Presenting this relationship accurately in instruction ensures that students appreciate the progressive and self-correcting nature of scientific understanding, aligning with the inquiry standards assessed in professional certification.

Experimentation, Variables, and Control

An experiment transforms abstract questions into testable procedures. Its design depends on the identification of independent, dependent, and controlled variables. The independent variable is deliberately manipulated, the dependent variable represents the measured outcome, and controlled variables are held constant to isolate causal relationships. This structure allows researchers to determine how one factor influences another while minimizing confounding effects. Replication strengthens validity by ensuring that results are not due to random variation. In educational environments, students often learn these concepts through guided inquiry labs, which emphasize systematic planning and data evaluation. Emphasizing reproducibility helps learners internalize that reliable conclusions arise from consistency across repeated trials. The proper use of control groups, blind tests, and standard procedures reflects professional ethics and intellectual rigor. Within the TExES framework, such understanding demonstrates mastery of scientific reasoning and process skills. The capacity to design and evaluate experiments also mirrors classroom practice, where teachers must translate abstract inquiry principles into accessible learning experiences. Through clear experimental logic, the structure of science becomes transparent and empowering for both educator and learner.

Data Analysis and Interpretation

Data analysis converts raw observations into meaningful conclusions. It involves organizing information, identifying patterns, and applying mathematical or statistical methods to discern relationships. Graphs, tables, and models aid visualization, allowing trends to be identified and anomalies to be scrutinized. Interpretation goes beyond numerical manipulation to integrate results within a theoretical context. This requires critical judgment—distinguishing between significant and coincidental findings, acknowledging error margins, and recognizing the influence of sample size and design limitations. The ethical dimension of interpretation demands that data never be altered to fit expectations. Transparency in methodology and honesty in reporting are central to the credibility of science. In the classroom, data analysis activities cultivate quantitative literacy and reinforce the concept that conclusions must emerge from evidence. On the TExES exam, such skills are assessed through questions that require reading graphs, evaluating correlations, and drawing reasoned inferences. Ultimately, interpretation represents the bridge between empirical work and conceptual understanding, ensuring that data illuminate rather than obscure the reality they describe.

Communication, Verification, and Peer Review

The scientific enterprise depends on the open exchange of knowledge. Communication allows results to be scrutinized, replicated, and integrated into the collective body of science. Reports, journals, presentations, and digital archives all serve as vehicles for dissemination. Effective communication requires precision of language, clarity of data representation, and acknowledgment of sources. Verification occurs when independent investigators reproduce findings under comparable conditions. Peer review formalizes this process, acting as the mechanism by which science polices its own credibility. It ensures that conclusions are grounded in sound methodology and logical reasoning rather than personal assertion. For educators and candidates alike, understanding the role of communication extends beyond publication; it encompasses ethical responsibility in representing information truthfully. Within the certification context, teachers are expected to model these values for students, demonstrating that collaboration and critique are strengths, not threats, to knowledge. The reliability of science ultimately depends on this culture of transparency, which transforms individual inquiry into collective advancement.

Ethics and Safety in Scientific Practice

Scientific inquiry operates within moral and social boundaries that safeguard integrity and human welfare. Ethical conduct requires honesty in data collection, respect for living organisms, environmental responsibility, and the responsible use of technology. Safety protocols protect both researchers and participants, reinforcing that precision and caution are inseparable. In classrooms and laboratories, teachers bear the responsibility of modeling proper safety behavior: wearing protective equipment, labeling substances, disposing of chemicals appropriately, and maintaining orderly environments. Ethical inquiry also includes intellectual transparency—crediting sources, avoiding plagiarism, and disclosing conflicts of interest. Violations of these principles undermine public trust and compromise the educational mission of science. Understanding ethics as a living framework rather than a list of prohibitions allows teachers to guide students toward conscientious citizenship. The TExES examination evaluates awareness of these responsibilities, emphasizing that science education involves character as well as cognition. Ethical inquiry ensures that the pursuit of knowledge contributes positively to human understanding and societal wellbeing.

Chapter 2 — Physics

Foundations of Physical Measurement and Units

Physics begins with the act of measuring and describing the natural world in quantitative terms. Every experiment or observation depends on a coherent system of units, standards, and definitions that ensure results are consistent and reproducible across time and place. The International System of Units (SI) provides the universal foundation, defining the meter, kilogram, and second as the base measures of length, mass, and time. Accuracy and precision are distinguished through calibration and the analysis of uncertainty. Dimensional analysis links different physical quantities, confirming that mathematical relationships remain consistent with the laws of nature. In physics, every measurable quantity—from velocity and energy to charge and temperature—can be traced to these base dimensions. The significance of measurement lies not only in establishing numerical value but in revealing the relationships between phenomena. Standards such as the speed of light or the cesium atomic clock define constants that ground modern science in stable reference points. Understanding these conventions allows scientists to communicate and verify data globally. Within the TExES framework, competence in measurement and unit conversion demonstrates mastery of the quantitative language through which physical theory is expressed and tested.

Motion in One and Two Dimensions

The study of motion forms the starting point for classical mechanics, uniting observation with mathematical representation. In one-dimensional motion, displacement, velocity, and acceleration describe how an object's position changes over time under uniform or varying forces. Equations of kinematics capture these relationships, enabling predictions of future positions or velocities when initial conditions are known. Extending motion to two dimensions introduces vector quantities, which express both magnitude and direction. Projectile motion exemplifies this by combining horizontal uniform motion with vertical acceleration due to gravity. The graphical and algebraic treatment of vectors allows scientists to decompose motion into components and reassemble them into coherent trajectories. The principle of independence of motion ensures that horizontal and vertical components act separately but combine to produce observed paths. In uniform circular motion, the acceleration is always directed toward the center, demonstrating that even constant-speed motion requires a continual change in direction—a fundamental insight linking force and acceleration. Understanding these laws provides a conceptual bridge between observation and Newtonian dynamics. Motion analysis is therefore not simply a topic within physics; it is the language through which all physical change can be interpreted and predicted.

Newtonian Dynamics and Forces

Isaac Newton's formulation of the laws of motion transformed human understanding of the universe by connecting forces to changes in motion. His first law defined inertia, the property of matter that resists changes in velocity. The second law expressed a precise quantitative relationship—force equals mass times acceleration—establishing that motion changes only when acted upon by a net external force. The third law introduced reciprocity: every action produces an equal and opposite reaction. Together, these principles explain the behavior of everything from a falling apple to planetary orbits. Forces may arise from contact interactions such as friction or tension, or from fields like gravitation and electromagnetism. The vector nature of force means that direction and magnitude jointly determine an object's acceleration. Equilibrium occurs when all forces cancel, producing no acceleration, a condition central to structural stability and engineering design. Newtonian dynamics remains the bedrock of classical mechanics, applicable to most everyday situations where velocities are small relative to the speed of light. Understanding these relationships allows teachers and students to connect abstract laws with the tangible mechanics of motion, forming the analytical foundation for every other branch of physical science.

Work, Energy, and Power

Energy provides a unifying concept that links diverse physical processes through the principle of conservation. Work is defined as the transfer of energy through force applied over a distance, integrating the ideas of motion and dynamics. When a constant force acts along the direction of motion, work equals the product of force and displacement; for varying forces, it is determined through the integral of force over distance. Kinetic energy quantifies the energy of motion, while potential energy reflects stored capacity due to position or configuration. The work–energy theorem connects these quantities, stating that the net work done on an object equals its change in kinetic energy. Power measures the rate at which work is performed or energy is transferred, expressed as energy per unit time. In real systems, friction and resistance transform mechanical energy into heat, but the total energy of a closed system remains constant. This conservation law underlies thermodynamics, electromagnetism, and quantum mechanics alike. Understanding energy flow enables students to see physics as an interconnected whole, where motion, force, and transformation merge into one comprehensive principle of natural order.

Momentum, Collisions, and Conservation Principles

Momentum expresses both the magnitude and direction of an object's motion, defined as the product of mass and velocity. Its significance lies in its conservation: within a closed system free of external forces, the total momentum remains constant. This law applies universally—from atomic interactions to the motion of galaxies—forming one of the most powerful predictive tools in physics. Collisions illustrate this principle vividly. In elastic collisions, both momentum and kinetic energy are conserved, while in inelastic collisions, momentum is preserved but energy transforms into heat or deformation. The concept of impulse, the product of force and time, links force to the change in momentum. These relationships form the basis for understanding mechanical stability, propulsion, and impact analysis. The center-of-mass framework extends conservation principles to systems of particles, allowing complex interactions to be reduced to simpler equivalent models. For educators, demonstrating conservation laws helps students recognize that while forms of motion may vary, the underlying quantities governing them obey universal symmetry. Momentum, like energy, serves as a fundamental invariant across all physical domains.

Rotational Motion and Equilibrium

Rotation introduces angular analogs to linear motion, expanding mechanics into a broader geometry of movement. Angular displacement, velocity, and acceleration describe how rigid bodies rotate about fixed axes, while torque represents the rotational equivalent of force. The relationship between torque, moment of inertia, and angular acceleration defines the rotational form of Newton's second law. Just as mass resists linear acceleration, moment of inertia resists angular change, depending on how mass is distributed relative to the axis of rotation. Work and energy also extend to rotational systems through rotational kinetic energy and power associated with angular motion. Static equilibrium occurs when both the net force and net torque on a body are zero, ensuring stability. Engineers rely on these principles in the design of structures, levers, and mechanical systems. Understanding balance, torque, and rotational dynamics allows prediction of motion in natural and human-made systems alike—from planetary spin to turbine mechanics. Within the TExES framework, mastery of rotational motion and equilibrium indicates deep comprehension of the symmetries and conservation laws that unify the study of mechanics.

Gravitation, Fluids, and Oscillatory Systems

Gravitation extends Newtonian dynamics from terrestrial motion to celestial mechanics. Newton's law of universal gravitation establishes that every mass attracts every other mass with a force proportional to their masses and inversely proportional to the square of their separation. This relationship explains planetary orbits, tides, and the structure of galaxies. Within Earth's environment, gravity determines weight and potential energy, linking macroscopic motion to universal laws. Fluid mechanics introduces the behavior of substances that flow—liquids and gases governed by pressure, buoyancy, and dynamic equilibrium. Bernoulli's principle describes how variations in velocity correspond to pressure differences, explaining lift, circulation, and flow. Oscillatory systems, such as pendulums and springs, exhibit periodic motion characterized by restoring forces proportional to displacement. These vibrations form the basis of wave motion, sound, and alternating electrical currents. Collectively, gravitation, fluids, and oscillations demonstrate how energy and force propagate through continuous media, uniting solid mechanics with thermodynamic and wave-based phenomena. Their study reinforces the concept that equilibrium and periodicity lie at the heart of natural stability and transformation.

Thermodynamics and Energy Transfer

Thermodynamics bridges microscopic motion with macroscopic behavior, describing how energy manifests as heat and work within systems. The zeroth law establishes thermal equilibrium, providing the basis for temperature measurement. The first law defines the conservation of energy in thermodynamic form: the change in internal energy equals the heat added to a system minus the work performed by it. The second law introduces entropy, quantifying the irreversibility of natural processes and defining the direction of energy flow. Engines, refrigerators, and biological systems all operate under these constraints, transforming energy within limits imposed by efficiency and entropy. The concept of thermal equilibrium links directly to the molecular theory of gases, where temperature corresponds to average kinetic energy. Understanding these principles provides insight into weather patterns, chemical reactions, and metabolic processes. In physics education, thermodynamics illustrates how the abstract conservation laws of mechanics extend into everyday experiences of heat and transformation. For certification candidates, it represents mastery of one of science's most integrative disciplines—connecting mechanics, chemistry, and environmental science within a single theoretical framework.

Chapter 3 — Chemistry

Chemical Foundations and Matter

Chemistry begins with the study of matter—anything that has mass and occupies space—and the transformations it undergoes. All observable materials are composed of atoms, the smallest units retaining the identity of elements. Atoms combine to form molecules and compounds through definite ratios governed by chemical laws. Matter exists in three primary states—solid, liquid, and gas—each defined by particle arrangement and energy. Understanding these forms provides a framework for explaining physical and chemical changes. Physical changes alter appearance without changing identity, while chemical changes involve the breaking and forming of bonds to produce new substances. Accurate measurement and classification of matter depend on the International System of Units, ensuring global consistency in scientific communication. From the atomic to the macroscopic level, matter obeys quantifiable rules, linking structure to properties. The study of chemistry therefore unites observation with theory, describing both how matter behaves and why it behaves that way. Within the TExES Science 7–12 framework, this foundational understanding supports every later exploration of bonding, reactions, and energy exchange.

Atoms, Molecules, and Atomic Theory

The atomic theory defines the structural basis of chemistry. Originating with Dalton's postulates, it asserts that all matter consists of indivisible atoms that combine in fixed proportions to form compounds. Each element contains identical atoms distinguished by mass and chemical behavior. Advances in experimental evidence—such as cathode ray studies, Rutherford's gold foil experiment, and quantum mechanics—revealed that atoms consist of nuclei containing protons and neutrons surrounded by electron clouds. The atomic number represents the number of protons, determining elemental identity, while isotopes differ in neutron count. Molecules form through the chemical bonding of atoms, governed by electrostatic attraction and energy minimization. Ions emerge when electrons are lost or gained, producing charged species central to reactivity. The periodic table arranges elements according to increasing atomic number, revealing recurring patterns in valence and bonding behavior. Understanding atomic structure allows prediction of molecular stability, polarity, and reaction types. In educational practice, these principles translate abstract atomic models into tangible explanations for real-world chemical phenomena, establishing the conceptual scaffolding of all chemical science.

Chemical Bonding and Molecular Structure

Bonding explains how atoms achieve stable configurations by sharing or transferring electrons. Ionic bonding involves electrostatic attraction between oppositely charged ions, typical of metals and nonmetals. Covalent bonding occurs when atoms share electrons, producing discrete molecules with defined geometries. Metallic bonding allows delocalized electrons to move freely among metal nuclei, accounting for conductivity and malleability. Molecular structure depends on electron-pair repulsion and orbital hybridization, described by valence shell electron pair repulsion (VSEPR) and molecular orbital theories. The spatial arrangement of atoms determines polarity, intermolecular forces, and physical properties such as boiling point or solubility. Bond enthalpy quantifies energy required to break bonds, linking molecular structure to thermochemical behavior. Understanding bonding requires visualizing both the electron density distribution and the three-dimensional geometry that results from atomic interactions. For examination purposes, mastery of this topic reflects an ability to connect symbolic representations—Lewis structures, formulas, and models—with the energetic and spatial realities they describe, thereby linking microscopic behavior to observable macroscopic properties.

Chemical Reactions and Stoichiometry

Chemical reactions are the central events of chemistry—processes in which reactants transform into products with new compositions and properties. Each reaction follows the law of conservation of mass: atoms are neither created nor destroyed but rearranged in new combinations. Balanced chemical equations express this conservation numerically, allowing calculation of quantities through stoichiometry. The mole concept provides the bridge between atomic scale and laboratory measurement, defining Avogadro's number as the link between individual particles and macroscopic amounts. Determining limiting reagents, theoretical yield, and percent yield ensures quantitative accuracy in both analysis and synthesis. Reactions occur in diverse forms—precipitation, redox, acid-base, and combustion—each driven by energy and electron transfer. Mastery of stoichiometric reasoning demonstrates the analytical precision required in scientific practice. In classrooms, it reinforces logical problem-solving and proportional thinking, skills tested rigorously within professional examinations. Stoichiometry transforms chemistry from descriptive observation to predictive science, empowering accurate control over material transformation.

States of Matter and Gas Behavior

The physical state of matter reflects the balance between kinetic energy and intermolecular forces. Gases, characterized by widely separated particles, obey relationships described by the ideal gas law: $PV = nRT$. Pressure, volume, temperature, and mole quantity interact predictably under this model. Real gases deviate at high pressures or low temperatures, where interactions and finite particle volumes become significant. Liquids maintain fixed volume but variable shape, with molecules bound by cohesive forces such as hydrogen bonding and van der Waals interactions. Solids possess definite shape and volume, organized into crystalline or amorphous structures. Transitions among these states involve energy changes that can be quantified through phase diagrams. Understanding gas laws and intermolecular behavior provides insight into atmospheric chemistry, solution formation, and thermodynamic equilibrium. It illustrates how molecular motion translates into measurable macroscopic properties—an essential link in the continuum of physical chemistry. For TExES candidates, proficiency in these laws signifies readiness to interpret and apply physical principles within chemical systems.

Thermochemistry and Energy Changes

Thermochemistry explores how energy accompanies chemical and physical changes. Every reaction involves either the absorption or release of heat, corresponding to endothermic or exothermic processes. The first law of thermodynamics establishes that energy is conserved: it may change form but cannot be created or destroyed. Enthalpy quantifies heat exchange at constant pressure, while calorimetry measures this transfer experimentally. Standard enthalpies of formation and reaction enable prediction of heat flow under given conditions. Hess's law demonstrates that overall enthalpy change depends only on initial and final states, not on the reaction path. Chemical stability and spontaneity arise from the interplay of enthalpy, entropy, and free energy. Thermochemical data therefore reveal not just how much energy changes but why certain reactions occur spontaneously. Understanding these relationships provides a bridge between microscopic bonding energy and macroscopic thermal behavior. It also connects chemistry to energy technology, environmental science, and materials design, illustrating the unity of natural and applied scientific inquiry.

Chemical Kinetics and Equilibrium

Kinetics examines how quickly reactions proceed and what factors influence their rates. The rate law expresses the relationship between concentration and time, while activation energy defines the minimum energy needed for effective collisions. Temperature, catalysts, and surface area alter reaction rates by changing collision frequency or energy distribution. Mechanistic studies identify intermediate steps that reveal how complex transformations unfold. Equilibrium describes the dynamic balance reached when forward and reverse reactions occur at equal rates. The equilibrium constant quantifies system position and predicts shifts caused by temperature, concentration, or pressure changes according to Le Châtelier's principle. Together, kinetics and equilibrium explain both the pace and extent of reactions. Their understanding allows control over industrial synthesis, biological regulation, and environmental processes. Within the TExES framework, mastery of these concepts signifies the ability to interpret dynamic systems as balanced yet responsive, embodying the self-regulating nature of chemical behavior in the physical world.

Acids, Bases, and Electrochemistry

Acid-base theory and electrochemistry represent two expressions of electron transfer. Acids donate protons, bases accept them, and their strength depends on ionization extent in water. The pH scale quantifies acidity, linking hydrogen ion concentration to logarithmic expression. Buffers maintain stable pH through conjugate acid-base pairs, essential for biological and environmental equilibrium. Electrochemistry extends this idea to oxidation-reduction reactions, where electron flow produces electrical energy or drives chemical change. Galvanic cells convert chemical energy into electricity, while electrolytic cells reverse this process through applied potential. Standard electrode potentials allow prediction of spontaneous reactions and voltage output. Together, these principles form the foundation for understanding corrosion, batteries, and biological energy transfer. For teachers and students alike, mastering acid-base and electrochemical processes integrates molecular theory with observable phenomena such as voltage, corrosion, and titration curves—key areas of both practical and theoretical competence in the science of chemistry.

Download Audio file and flashcards

Chapter 4 — Biology

The Nature and Organization of Living Systems

Life is characterized by organization, regulation, and the capacity for growth and adaptation. All living systems share fundamental properties: they are composed of cells, maintain homeostasis, respond to environmental stimuli, utilize energy, and reproduce. Biological organization extends from molecular components to complex multicellular organisms, forming hierarchies of structure and function. At the molecular level, carbon-based compounds enable chemical reactions essential for life. At the cellular level, compartmentalization allows specialized processes to occur efficiently. In multicellular organisms, tissues, organs, and systems coordinate through signaling and feedback mechanisms. The study of biology thus explores both the unity and diversity of life, seeking to explain how order emerges from molecular interactions. Organisms persist through dynamic equilibrium—balancing internal conditions with external change. This self-regulating capacity distinguishes living matter from nonliving systems. For future educators and scientists, understanding biological organization provides the conceptual foundation for connecting microscopic mechanisms to ecological and evolutionary patterns that sustain life on Earth.

The Chemistry of Life and Macromolecular Structure

The chemistry of life is rooted in the interactions of atoms that form biological macromolecules. Carbon's ability to form four covalent bonds allows for immense structural diversity, enabling the creation of carbohydrates, lipids, proteins, and nucleic acids. Water, with its polarity and hydrogen-bonding properties, serves as the universal solvent and regulator of biochemical reactions. Carbohydrates supply immediate energy and structural support; lipids store energy and form membranes; proteins perform catalytic and regulatory roles; nucleic acids encode and transmit genetic information. Each macromolecule's structure dictates its function—folding patterns, charge distribution, and molecular recognition determine biological activity. Enzymes lower activation energy, controlling metabolic pathways with remarkable specificity. The interplay between structure and function at the molecular level defines the essence of biochemistry and cellular physiology. Recognizing how chemical principles underpin biological processes allows educators to bridge the gap between physics, chemistry, and life sciences, reflecting the interdisciplinary scope of modern biology as assessed by the TExES framework.

Cell Structure and Function in Eukaryotes and Prokaryotes

The cell represents the fundamental unit of life, displaying both autonomy and cooperation within multicellular organisms. Prokaryotic cells, including bacteria and archaea, lack membrane-bound organelles but possess remarkable biochemical versatility. Eukaryotic cells exhibit compartmentalization through specialized organelles such as the nucleus, mitochondria, endoplasmic reticulum, and Golgi apparatus. These structures coordinate energy production, protein synthesis, and molecular transport. The cytoskeleton provides mechanical support and facilitates movement, while the plasma membrane governs the selective exchange of materials. Cellular diversity arises from structural adaptations reflecting function—nerve cells transmit signals, muscle cells contract, and plant cells maintain rigidity through cell walls and vacuoles. Understanding cellular organization allows insight into how structure determines physiological role. At its core, cell theory unites all life under a single principle: every living organism is composed of one or more cells, and all cells arise from preexisting cells. This continuity of cellular life underscores both evolutionary connection and biological unity.

Cellular Membranes and Transport Mechanisms

Biological membranes form the interface between a cell and its environment, maintaining distinct internal conditions essential for life. Composed primarily of phospholipids and proteins, membranes exhibit fluidity and selective permeability. This arrangement enables dynamic control over molecular movement through diffusion, osmosis, and active transport. Channel and carrier proteins facilitate the passage of ions and nutrients, while pumps use energy to move substances against concentration gradients. Endocytosis and exocytosis allow the exchange of large molecules, maintaining communication with the extracellular space. Membrane receptors detect chemical signals, initiating cascades that regulate cellular activity. The fluid mosaic model describes the dynamic nature of membrane structure, where components move laterally within the lipid bilayer. Understanding transport mechanisms reveals how cells maintain homeostasis, respond to stimuli, and coordinate metabolic function. In education and research, the study of membranes demonstrates how biological systems balance stability and change, integrating chemistry, physics, and biology into a coherent model of living regulation.

Energy Transformations and Metabolism

Metabolism encompasses all chemical reactions that sustain life, converting energy from one form to another. Cells extract energy through catabolic pathways that break down molecules, and they use anabolic pathways to synthesize complex compounds. The universal energy currency of the cell, adenosine triphosphate (ATP), stores and releases energy through the phosphorylation of substrates. Cellular respiration transforms glucose and oxygen into carbon dioxide, water, and ATP through glycolysis, the Krebs cycle, and oxidative phosphorylation. In contrast, photosynthetic organisms convert light energy into chemical energy using chlorophyll and pigment complexes in chloroplasts. Enzymatic regulation ensures efficiency, preventing wasteful reactions through feedback inhibition. Metabolism reflects the unity of life's chemical design: diverse organisms rely on the same fundamental principles of energy capture and transformation. In the context of the TExES framework, comprehension of metabolism demonstrates an understanding of how biochemical energy drives cellular work and sustains biological organization across all levels of complexity.

DNA, Genes, and Chromosomal Organization

Deoxyribonucleic acid (DNA) embodies the molecular blueprint of life. Its double-helical structure, composed of nucleotide pairs joined by hydrogen bonds, provides stability and fidelity in information storage. Each gene represents a sequence of nucleotides coding for a specific protein or functional RNA. DNA is organized into chromosomes, compacted by histone proteins in eukaryotes and arranged circularly in most prokaryotes. During replication, complementary base pairing ensures accurate duplication, allowing genetic information to pass from cell to cell and generation to generation. Mutations—changes in nucleotide sequence—serve as both potential errors and the raw material of evolution. Understanding chromosomal organization reveals how genetic material is maintained, expressed, and inherited. Molecular biology unites structure with function, showing that genetic information is not static but dynamically regulated. In educational practice, DNA's elegant simplicity exemplifies how complexity arises from the precise arrangement of simple components, demonstrating the informational logic that underlies all biological processes.

Gene Expression, Regulation, and Protein Synthesis

The flow of genetic information follows the central dogma: DNA is transcribed into RNA, which is translated into protein. Transcription produces messenger RNA using one DNA strand as a template, a process controlled by promoters, enhancers, and regulatory proteins. Ribosomes then translate the mRNA sequence into a polypeptide chain, interpreting each triplet codon into its corresponding amino acid. Post-translational modifications fold and activate proteins, determining their cellular role. Regulation occurs at multiple levels—transcriptional, translational, and post-translational—allowing cells to respond to environmental and developmental signals. In eukaryotes, the compartmentalization of transcription in the nucleus and translation in the cytoplasm introduces additional layers of control. This complexity ensures that gene expression is precise yet adaptable. Understanding these mechanisms illuminates how genotype produces phenotype and how cellular differentiation arises from a single genetic code. In modern biology, gene regulation exemplifies the orchestration of molecular processes that define life's versatility and responsiveness.

Cell Division, Growth, and Reproduction

Cell division enables growth, repair, and reproduction in living systems. The cell cycle comprises interphase—where DNA replicates and the cell prepares for division—and the mitotic or meiotic phases, during which genetic material is distributed. Mitosis results in two identical daughter cells, maintaining chromosome number for somatic functions. Meiosis, by contrast, halves the chromosome number to produce gametes, ensuring genetic diversity through crossing-over and independent assortment. Regulation of the cell cycle is tightly controlled by checkpoints and signaling proteins that prevent uncontrolled division. Errors in these controls may lead to diseases such as cancer, underscoring the necessity of molecular precision. In multicellular organisms, coordinated cell division maintains tissue integrity and organismal growth. Understanding mitosis and meiosis reveals the continuity of life through successive generations and the balance between stability and variation that underpins evolutionary change.

Principles of Inheritance and Molecular Genetics

Inheritance explains how traits are transmitted from parents to offspring according to predictable patterns. Mendel's experiments with pea plants established the laws of segregation and independent assortment, demonstrating that discrete units of heredity determine phenotype. Modern genetics identifies these units as genes located on chromosomes. Variations arise from mutations, recombination, and the random fusion of gametes. Molecular genetics extends classical principles by showing how allelic differences influence protein function and how gene interactions produce complex traits. Linkage analysis and mapping have revealed that genes may be inherited together if located near one another on the same chromosome. In the era of genomics, the scope of inheritance now encompasses entire DNA sequences, enabling the study of populations and evolutionary lineages. Understanding inheritance provides a bridge between molecular biology and evolution, illustrating how continuity and change coexist within the fabric of life.

Evolutionary Theory and Mechanisms of Change

Evolution explains the diversity of life through the process of descent with modification. Natural selection acts on heritable variation, favoring traits that enhance survival and reproduction. Over time, advantageous genes become more frequent in populations, leading to adaptation and speciation. Additional mechanisms—mutation, gene flow, genetic drift, and recombination—introduce variation and alter allele frequencies. Fossil evidence, comparative anatomy, and molecular data all confirm the shared ancestry of organisms. Evolution operates at multiple scales, from microevolutionary changes within populations to macroevolutionary patterns observed across geological time. The theory of evolution unites all biological disciplines under a single explanatory framework, providing coherence to the study of life. For educators, presenting evolution as a scientific process based on evidence cultivates analytical thinking and underscores the dynamic nature of biological knowledge, in line with the inquiry-based standards of professional certification.

Biological Classification and Diversity of Life

The diversity of living organisms is organized through a hierarchical system of classification known as taxonomy. Modern taxonomy uses morphological, genetic, and molecular data to group organisms into domains, kingdoms, phyla, classes, orders, families, genera, and species. The three-domain system—Bacteria, Archaea, and Eukarya—reflects fundamental differences in cellular structure and genetic organization. Within this hierarchy, phylogenetics reconstructs evolutionary relationships using DNA sequences and molecular markers. Biodiversity encompasses variation within and among species, forming the foundation of ecological resilience and evolutionary potential. Understanding classification allows students to appreciate both unity and differentiation within the living world. It provides a systematic language for describing the immense variety of life and situates individual species within the larger continuum of biological history. This organization mirrors the conceptual order that underpins all scientific reasoning, emphasizing structure, relationship, and evidence-based categorization.

Ecology, Populations, and Interdependence of Organisms

Ecology examines how organisms interact with each other and their environment, forming dynamic systems of energy and matter exchange. Populations consist of individuals of the same species in a defined area, regulated by birth rates, death rates, and resource availability. Communities arise when multiple populations coexist, connected through food webs and ecological niches. Energy flows through ecosystems via producers, consumers, and decomposers, while nutrients cycle through biogeochemical processes. Environmental factors—climate, competition, and human activity—shape population dynamics and ecosystem stability. Interdependence ensures that no organism exists in isolation; survival depends on cooperation, competition, and adaptation. At a global scale, ecology reveals how biological and physical systems sustain the biosphere. For future science educators, understanding ecological principles is vital for fostering environmental literacy and stewardship, aligning biological study with ethical responsibility toward Earth's interconnected systems.

Chapter 5 — Heredity and Evolution of Life

Foundations of Genetic Continuity

Heredity ensures that organisms transmit essential traits across generations through the stability of genetic material. The mechanisms of inheritance explain both the constancy and variation seen in life forms. Within every cell, DNA encodes the instructions for structure, function, and reproduction. During cell division, replication preserves genetic identity, while recombination introduces diversity. Heredity reflects a delicate balance between fidelity and change—stability ensures species continuity, but variation fuels evolution. Early theories of blending inheritance were replaced by Mendelian principles, which demonstrated that genes remain discrete units. Modern genetics integrates these classical foundations with molecular evidence, revealing how genes are expressed and regulated. Through this integration, heredity is not a static process but a dynamic system of information transfer, shaping development, adaptation, and survival. In educational practice, understanding genetic continuity provides a conceptual bridge linking cellular biology, molecular mechanisms, and evolutionary outcomes that define modern scientific thought.

Mendelian Principles of Inheritance

Gregor Mendel's experiments with pea plants introduced a quantitative understanding of heredity that underpins all modern biology. By controlling pollination and observing patterns across generations, he discovered that traits are determined by paired hereditary factors—now known as genes—that segregate independently during gamete formation. His Law of Segregation describes how alleles separate during meiosis, ensuring each gamete carries only one factor for each trait. The Law of Independent Assortment extends this by showing how genes for different traits are inherited independently, provided they reside on separate chromosomes. Mendel's ratios, confirmed through statistical analysis, established predictability in inheritance. These insights transformed biology into a science governed by mathematical principles, bridging observation with theory. The enduring relevance of Mendel's work lies in its predictive power—allowing scientists to model inheritance patterns across organisms and explore the molecular explanations that later emerged with chromosome theory.

Molecular Basis of Heredity

At the molecular level, heredity resides in DNA, a double-helical polymer that encodes genetic information through the sequence of nucleotide bases. The discovery of the DNA structure revealed how replication ensures accurate transmission of hereditary material. Complementary base pairing allows one strand to serve as a template for the other, maintaining genetic fidelity. Within the genome, genes represent discrete sequences that code for proteins or functional RNA molecules. Transcription converts DNA into messenger RNA, while translation decodes RNA into polypeptide chains using ribosomal machinery. The universality of the genetic code reflects evolutionary conservation among all living systems. Mutations, though rare, introduce variations that may alter protein structure and function, providing material for natural selection. The molecular understanding of heredity thus connects chemistry to evolution, showing that life's complexity emerges from fundamental biochemical logic.

Chromosomal Theory and Genetic Mapping

The chromosomal theory of inheritance integrated Mendel's laws with cytological evidence, establishing chromosomes as carriers of genes. During meiosis, homologous chromosomes segregate and assort independently, paralleling Mendel's predictions. Each gene occupies a specific locus on a chromosome, and linkage occurs when genes reside close together, reducing recombination frequency. Crossing over during prophase I introduces new combinations, creating genetic maps that measure distances based on recombination percentages. Early mapping studies in fruit flies provided empirical proof of this mechanism, revealing that physical exchange of chromosomal segments corresponds to measurable inheritance patterns. Modern mapping integrates classical data with molecular markers, such as restriction fragment length polymorphisms and sequence-tagged sites. These methods enable precise localization of genes associated with inherited conditions. Understanding chromosomal behavior links classical genetics with modern genomics, allowing educators and students to visualize inheritance as a structural and spatial phenomenon rather than an abstract concept.

Gene Regulation and Expression

Genes function dynamically rather than statically, turning on or off in response to developmental and environmental cues. In prokaryotes, regulation often involves operon systems, such as the lac and trp operons, which control enzyme synthesis through feedback and repression. In eukaryotes, regulation is more intricate, involving chromatin remodeling, transcription factors, enhancers, and epigenetic modification. RNA splicing, transport, and degradation further modulate expression, producing tissue-specific patterns from a shared genome. The regulation of gene expression explains differentiation, homeostasis, and adaptation. It also reveals how disruptions in control—whether through mutation or environmental interference—lead to disease. By studying regulatory systems, learners grasp how a single genome generates immense biological diversity. For exam preparation, mastery of these principles clarifies the continuity between genetic information and observable phenotype, demonstrating how living organisms coordinate molecular processes with environmental demands.

Mutations, Variation, and Genetic Engineering

Mutations represent the raw material for evolution, introducing new alleles into populations. They occur through errors in replication, recombination, or exposure to mutagens. While many mutations are neutral or deleterious, some confer adaptive advantages. Modern biotechnology has harnessed the mechanisms of mutation and recombination for controlled manipulation. Recombinant DNA technology allows scientists to clone, modify, and transfer genes across species, producing genetically engineered crops, therapeutic proteins, and model organisms. Techniques such as polymerase chain reaction and DNA sequencing provide unprecedented control over genetic analysis. These tools bridge molecular biology and applied science, enabling precise exploration of heredity. In an educational context, understanding mutation and engineering links conceptual knowledge with technological relevance, highlighting both the creative potential and ethical responsibility inherent in manipulating genetic information.

Patterns of Inheritance Beyond Mendel

While Mendel's principles established foundational laws, real-world genetics reveals complexity beyond simple dominance. Incomplete dominance produces intermediate phenotypes, while codominance allows both alleles to express equally. Polygenic inheritance governs traits influenced by multiple genes, such as height or skin color, displaying continuous variation. Environmental factors further modify expression, emphasizing the interaction between genotype and environment. Epistasis occurs when one gene masks the expression of another, adding additional layers of complexity. Non-nuclear inheritance, such as mitochondrial DNA transmission, follows unique maternal patterns. The study of these variations enriches understanding of biological diversity and reveals the limits of classical models. For educators, this complexity reinforces that heredity operates as a spectrum of influences rather than a set of rigid categories, preparing students for more advanced molecular and evolutionary study.

Population Genetics and Evolutionary Mechanisms

Evolution operates on populations rather than individuals, manifesting as changes in allele frequencies over time. The Hardy–Weinberg equilibrium provides a mathematical model describing stable populations where allele frequencies remain constant absent evolutionary forces. In practice, deviations from equilibrium reveal the influence of mutation, gene flow, genetic drift, nonrandom mating, and natural selection. Population genetics quantifies how these mechanisms shape genetic structure, forming the empirical basis of evolutionary theory. The synthesis of Mendelian principles with Darwinian selection created the modern understanding of evolution as a genetic process. Population-level analysis allows scientists to predict evolutionary outcomes and assess genetic health in species conservation. Within the TExES framework, comprehension of these dynamics equips educators to connect mathematical reasoning with biological processes, illustrating evolution as an inevitable consequence of heredity interacting with environment.

Speciation and Macroevolutionary Change

Speciation represents the branching of evolutionary lineages through reproductive isolation. Geographic, ecological, and behavioral barriers prevent gene flow, leading to genetic divergence. Over time, accumulated mutations and selection pressures establish new species adapted to distinct niches. The study of macroevolution extends beyond populations to patterns in the fossil record and comparative genomics, revealing continuity among all life forms. Molecular phylogenetics reconstructs evolutionary relationships by comparing DNA sequences, confirming descent with modification at the genomic level. Speciation thus bridges microevolutionary processes and large-scale biological diversity. Understanding these mechanisms provides a framework for explaining both the unity and variety of life, emphasizing that evolution is not a theory of replacement but of continual transformation and adaptation.

Evolutionary Evidence and the Unity of Life

Multiple lines of evidence converge to support evolutionary theory. Fossil sequences reveal transitional forms linking ancestral and modern species. Comparative anatomy demonstrates homologous structures shaped by divergent evolution. Molecular similarities—shared genetic codes, conserved proteins, and universal metabolic pathways—affirm common ancestry. Embryological parallels highlight developmental unity, while biogeography shows how isolation shapes diversity. Evolutionary processes explain both adaptation and imperfection, revealing that natural selection works through modification of existing structures. The coherence of these findings across disciplines underscores evolution as the unifying principle of biology. For teachers and students, integrating these evidences fosters a holistic understanding of science as an interconnected system of inquiry grounded in observation, experimentation, and synthesis.

Chapter 6 — Earth and Space Science

The Dynamic Earth System

Earth functions as an interconnected system composed of the geosphere, hydrosphere, atmosphere, and biosphere. Energy from the Sun and Earth's interior drives constant change within and among these spheres. Geological and climatic processes shape the surface while biological activity influences atmospheric composition and soil formation. Feedback mechanisms maintain stability through cycles of energy and matter, such as the carbon and water cycles. Understanding Earth as a dynamic system emphasizes how human actions can disrupt equilibrium, altering temperature, weather patterns, and ecosystems. Within scientific education, this systemic perspective fosters appreciation for planetary resilience and the delicate balance sustaining life.

Earth's Structure and Composition

Beneath Earth's surface lies a layered structure composed of crust, mantle, outer core, and inner core. The crust forms a thin, brittle shell, while the mantle's slow convection drives plate motion. The metallic core generates Earth's magnetic field, shielding life from solar radiation. By studying seismic waves, scientists infer internal composition and density variations. Minerals crystallize under specific temperature and pressure conditions, forming igneous, sedimentary, and metamorphic rocks that record geological history. Understanding these layers allows prediction of volcanic activity, earthquake behavior, and resource distribution—fundamental aspects of Earth science and environmental stewardship.

Plate Tectonics and the Changing Surface

The theory of plate tectonics unifies geology by explaining the movement of Earth's lithospheric plates atop the viscous asthenosphere. Driven by mantle convection and gravitational forces, these plates diverge, converge, or slide past each other, forming mountains, ocean basins, and fault systems. Continental drift reshapes global geography over millions of years, influencing ocean circulation and climate. Subduction recycles crust into the mantle, releasing energy that drives earthquakes and volcanism. Tectonic processes reveal the planet's internal dynamism and illustrate how gradual motion produces profound transformation—a concept central to interpreting Earth's past and forecasting its future.

The Rock Cycle and Geological Time

Rocks continually transform through the rock cycle—a sequence linking igneous formation, sedimentation, metamorphism, and melting. This cycle embodies Earth's regenerative capacity, where destruction and creation coexist. Geological time extends across billions of years, deciphered through radiometric dating and fossil evidence. The principle of uniformitarianism—processes observed today acting through immense time—allows reconstruction of planetary history. Stratigraphy records environmental shifts, mass extinctions, and continental movements. Understanding deep time broadens perspective on Earth's resilience and humanity's brief presence within its vast chronology.

Atmosphere and Climate Regulation

Earth's atmosphere, a blend of nitrogen, oxygen, and trace gases, moderates climate through radiative balance. Solar energy absorbed and re-emitted by land, oceans, and air drives weather and global circulation. Greenhouse gases trap infrared radiation, sustaining temperatures compatible with life, but human activities amplify this effect, altering climate stability. Atmospheric layers—troposphere to exosphere—exhibit gradients in temperature and composition that influence weather dynamics and protection from cosmic radiation. Climate regulation depends on complex interactions among energy, water vapor, and biospheric feedback, highlighting the delicate equilibrium sustaining planetary habitability.

The Hydrosphere and Ocean Systems

Water cycles continuously through evaporation, condensation, precipitation, and runoff, connecting all Earth systems. Oceans, covering more than two-thirds of the surface, regulate temperature and store vast amounts of energy. Currents driven by wind, salinity, and density distribute heat globally, influencing climate and ecosystems. The cryosphere—glaciers and polar ice—acts as both reservoir and climate indicator. Human reliance on freshwater resources underscores the importance of understanding hydrological processes. Ocean and water system studies reveal how planetary energy balance and life's distribution depend fundamentally on the properties of water.

Earth's Energy Balance and the Sun

The Sun supplies nearly all external energy driving atmospheric and biological processes. Solar radiation powers photosynthesis, fuels weather, and defines the climate zones of Earth. Variation in solar output and orbital geometry influences long-term cycles such as glaciations. Earth's internal energy, derived from radioactive decay and residual heat, powers plate tectonics and volcanism. Balancing incoming and outgoing radiation determines average temperature; disturbances lead to warming or cooling trends. This energy equilibrium explains both the stability and variability observed throughout geological and climatic history, linking astronomy and geology within a single continuum.

Weather, Oceans, and Global Circulation

Atmospheric motion results from uneven solar heating across Earth's surface. Air masses rise, cool, and condense, forming winds, clouds, and storms that distribute energy. The Coriolis effect shapes global wind patterns, driving ocean currents that moderate coastal climates. Large-scale systems such as El Niño and monsoons demonstrate the interconnectedness of atmosphere and ocean. Predicting weather relies on interpreting thermodynamic and dynamic processes through satellite observation and modeling. Understanding circulation reveals the planet as a dynamic engine balancing energy and matter, with implications for agriculture, navigation, and disaster preparedness.

Earth's History and Fossil Record

Earth's geological record chronicles biological and environmental evolution. Fossils embedded in sedimentary layers reveal the succession of life forms, extinction events, and climate transitions. Major eras—Precambrian, Paleozoic, Mesozoic, and Cenozoic—mark turning points in biological innovation and mass extinction. Continental rearrangements and volcanic activity often coincide with these transitions, illustrating the link between geology and evolution. Radiometric dating provides absolute ages, transforming paleontology into a quantitative science. By studying the fossil record, scientists reconstruct both the tempo and mode of evolutionary change, connecting deep Earth processes with the story of life itself.

Earth in the Solar System

Earth occupies a unique position in the solar system, orbiting within the habitable zone where liquid water persists. Its atmosphere, magnetic field, and geological activity distinguish it from neighboring planets. The solar system's formation from a rotating nebula explains planetary differentiation by mass and composition. Comparative planetology shows how Venus, Mars, and Earth evolved divergently under similar beginnings. Understanding Earth's celestial context enhances comprehension of its origin, stability, and potential vulnerability, emphasizing that planetary science extends beyond geology to include the physics of cosmic evolution.

The Universe and Cosmic Perspective

Beyond Earth, the universe expands through vast scales of space and time. Stars form within nebulae, fuse hydrogen into heavier elements, and eventually recycle material into new generations of stars and planets. Galaxies, composed of billions of stars, illustrate structure on the grandest scale, while cosmology seeks to explain origins through models such as the Big Bang. Studying the universe provides perspective on humanity's place within cosmic evolution. The same physical laws governing atoms apply to galaxies, affirming the unity of natural processes across all scales of existence.

Chapter 7 — Environmental Systems and Sustainability

Ecosystem Dynamics and Energy Flow

Ecosystems are complex networks of organisms interacting with each other and their physical environment through the transfer of energy and matter. Energy enters primarily through sunlight, captured by photosynthetic organisms that convert radiant energy into chemical form. This energy flows through trophic levels—producers, consumers, and decomposers—forming interconnected food webs. At each transfer, some energy dissipates as heat, following the laws of thermodynamics. Matter, unlike energy, cycles continuously through biological and geochemical pathways, maintaining ecosystem structure and function. Stability arises from feedback mechanisms that balance production and consumption, but disturbances can trigger succession and transformation. Biodiversity enhances resilience, allowing ecosystems to recover from natural or anthropogenic disruption. Understanding energy flow clarifies how life depends on the constant renewal of solar input and material recycling. For environmental scientists and educators, the concept of dynamic equilibrium illustrates both the fragility and persistence of natural systems, grounding sustainability in ecological balance rather than static preservation.

Nutrient Cycles and Biogeochemical Interactions

The cycling of nutrients—carbon, nitrogen, phosphorus, sulfur, and water—links biological activity with geological and atmospheric processes. The carbon cycle regulates climate by controlling atmospheric carbon dioxide levels through photosynthesis, respiration, and oceanic exchange. Nitrogen, essential for amino acids and nucleic acids, cycles through fixation, nitrification, and denitrification, often mediated by microorganisms. Phosphorus moves more slowly through rocks and soils, limiting productivity in many ecosystems. Human activity has accelerated these cycles, increasing nutrient runoff, eutrophication, and greenhouse gas emissions. Biogeochemical interactions reveal that no ecosystem functions in isolation; each depends on the global exchange of elements through land, water, and air. Disrupting these cycles alters ecosystem health and planetary stability. A deep understanding of nutrient cycling underscores sustainability as the art of maintaining flow without accumulation or depletion, emphasizing harmony between natural rates of renewal and human demand for resources.

Human Impacts on Natural Systems

Human expansion has become the dominant force shaping Earth's surface and atmosphere. Deforestation, industrialization, and agriculture modify landscapes, reducing biodiversity and altering water and carbon cycles. Habitat fragmentation isolates populations, leading to genetic erosion and ecosystem decline. Pollution from fossil fuel combustion contributes to acid rain and atmospheric instability, while overfishing and soil degradation deplete renewable resources faster than they can regenerate. These cumulative pressures define the Anthropocene—a period in which human activity rivals natural forces in global impact. Yet awareness of these consequences also provides the foundation for restoration and policy intervention. Environmental science emphasizes that sustainability depends not on halting progress but on integrating ecological limits into technological and economic systems. Recognizing humanity as both participant and steward within natural systems reframes environmental responsibility as a question of coexistence, adaptation, and long-term planetary resilience.

Pollution, Waste, and Environmental Health

Pollution arises when natural systems become overloaded with substances or energy beyond their capacity to absorb or neutralize. Air pollution from industrial emissions releases particulate matter, nitrogen oxides, and sulfur compounds that contribute to respiratory illness and climate forcing. Water pollution, stemming from agricultural runoff, plastics, and untreated waste, disrupts aquatic ecosystems and human health. Soil contamination reduces fertility and bioavailability of nutrients, often persisting for decades. The management of solid waste illustrates the tension between consumption and sustainability: landfills occupy space while decomposition releases methane. Advances in recycling and waste-to-energy technologies mitigate these impacts but require systemic cultural and economic change. Environmental health links ecosystem integrity with human well-being, showing that toxins and pathogens move freely through air, water, and food. Understanding pollution emphasizes prevention over remediation—an ethic central to sustainable development and global environmental stewardship.

Climate Systems and Global Change

Climate emerges from the balance of energy received from the Sun and emitted back into space. Greenhouse gases such as carbon dioxide, methane, and water vapor regulate this balance by trapping heat within the atmosphere. Human-induced increases in these gases enhance the greenhouse effect, driving global warming and altering precipitation, ocean currents, and storm frequency. Climate systems operate through feedbacks: melting ice reduces albedo, amplifying warming, while increased vegetation may enhance carbon sequestration. Observations of rising sea levels, shifting biomes, and species migrations confirm large-scale transformation. Climate change is not solely a physical phenomenon but a societal challenge linking science, policy, and ethics. Adaptation and mitigation strategies—renewable energy, reforestation, and carbon capture—illustrate attempts to align human activity with planetary limits. Understanding climate systems equips educators and citizens to interpret environmental data critically and act responsibly within a changing world.

Conservation Biology and Resource Management

Conservation biology seeks to preserve the diversity of life by understanding ecological processes that maintain species and habitats. It integrates genetics, population dynamics, and landscape ecology to identify critical thresholds for survival. Habitat corridors, protected areas, and sustainable harvest strategies are tools for balancing human use with ecological integrity. Resource management extends these principles to forests, fisheries, and mineral reserves, emphasizing renewable cycles and long-term planning. Overexploitation threatens both biodiversity and human welfare by destabilizing food webs and ecosystem services. Conservation emphasizes prevention, recognizing that extinction is irreversible. Through education and policy, the principles of conservation biology transform ethics into practice, fostering respect for life's interconnectedness. Sustainability in this context means ensuring that natural capital—biodiversity, soil, water, and atmosphere—remains sufficient to support future generations without diminishing the richness of the present biosphere.

Sustainable Agriculture and Food Systems

Agriculture sustains human civilization but also imposes significant environmental costs. Conventional practices relying on monocultures, chemical fertilizers, and pesticides often degrade soil, reduce biodiversity, and contaminate water. Sustainable agriculture seeks to harmonize productivity with ecological health through crop rotation, organic amendments, and integrated pest management. Soil conservation techniques such as contour plowing and agroforestry maintain fertility while reducing erosion. Advances in biotechnology and precision farming offer potential efficiency gains but must be guided by ethical and environmental considerations. Food systems also involve social and economic dimensions—equitable distribution, local production, and reduced waste contribute to global sustainability. Understanding agricultural sustainability reveals that feeding humanity depends not solely on technology but on ecological literacy and cultural adaptation. The goal is to produce food that nourishes people while regenerating the land, embodying the principle that human prosperity and planetary health are inseparable.

Renewable Energy and Technological Transitions

Energy production lies at the heart of sustainability. The transition from fossil fuels to renewable sources—solar, wind, geothermal, and hydroelectric—reduces greenhouse gas emissions and dependence on finite resources. Each technology presents unique advantages and challenges: solar energy offers decentralized access; wind turbines provide scalable capacity; geothermal systems yield constant output; and hydropower integrates storage with generation. Efficiency improvements in transportation and building design complement these shifts. However, infrastructure change requires policy support, investment, and public acceptance. The pursuit of renewable energy is not merely technological but cultural, redefining progress as balance rather than expansion. By harnessing natural energy flows instead of depleting reserves, humanity aligns economic development with planetary boundaries. Understanding this transition illustrates that sustainability emerges from innovation guided by ecological wisdom rather than exploitation.

Urbanization, Land Use, and Environmental Planning

Urban growth reshapes landscapes, concentrating populations and resources while generating waste, pollution, and social challenges. Effective environmental planning integrates green spaces, efficient transportation, and water management to reduce ecological footprints. Smart cities employ technology to monitor energy use, optimize traffic, and manage waste sustainably. Urban ecosystems can support biodiversity through green roofs, urban forests, and restored wetlands that improve air and water quality. Land use planning balances residential, industrial, and agricultural needs within regional ecological limits. The challenge lies in reconciling economic growth with environmental preservation, ensuring that urbanization enhances rather than diminishes quality of life. Environmental planning thus becomes both science and art—designing human habitats that coexist harmoniously with natural systems, exemplifying sustainability in practice.

Environmental Ethics, Policy, and Future Pathways

Sustainability ultimately rests on ethical foundations. Environmental ethics expands moral consideration beyond humanity to include all living systems and future generations. Policy translates these values into action through laws regulating pollution, resource use, and conservation. International agreements, such as those addressing climate and biodiversity, demonstrate that environmental stewardship transcends national boundaries. Education plays a pivotal role in shaping awareness and responsibility, cultivating the capacity to make informed, value-driven decisions. The future of sustainability depends on integrating science, culture, and policy into a shared vision of planetary well-being. Rather than seeking dominion over nature, humanity must embrace partnership with it. The ethic of sustainability calls for humility, foresight, and cooperation—a recognition that the prosperity of civilization is bound to the health of Earth's living systems.

Chapter 8 — Scientific Inquiry, Methodology, and Instruction

Nature and Purpose of Scientific Inquiry

Scientific inquiry is the systematic pursuit of understanding the natural world through observation, experimentation, and logical reasoning. It differs from casual observation by its commitment to objectivity, repeatability, and the critical evaluation of evidence. The goal of inquiry is not merely the accumulation of facts but the construction of coherent explanations that can be tested and refined. Scientists approach phenomena with curiosity tempered by skepticism, recognizing that every conclusion remains provisional and open to revision. This self-correcting nature gives science its reliability, allowing knowledge to evolve as new data emerge. Inquiry begins with questions derived from observation, leading to hypotheses that can be tested under controlled conditions. Data gathered through measurement and analysis provide the foundation for theory, while peer review ensures credibility. For educators, fostering inquiry means teaching students how to think scientifically—how to ask meaningful questions, gather evidence, and evaluate results critically. Inquiry thus becomes both a process of discovery and a disciplined way of knowing that extends beyond the laboratory into everyday reasoning and decision-making.

The Scientific Method and Experimental Design

The scientific method provides a logical framework for exploring questions about the natural world. It begins with observation and questioning, followed by the formulation of hypotheses—tentative explanations that can be tested through experiment or further observation. Effective experimental design identifies independent, dependent, and controlled variables, ensuring that results reflect causal relationships rather than coincidence. Replication and randomization enhance reliability, while control groups establish baselines for comparison. Quantitative measurements and precise documentation allow for reproducibility, a cornerstone of scientific validity. The process culminates in the analysis and interpretation of data to accept, modify, or reject the hypothesis. However, the method is not linear; it is iterative, adjusting as evidence accumulates. In educational settings, guiding students through the design of their own investigations fosters creativity and ownership of learning. Understanding experimental design empowers learners to distinguish between scientific evidence and anecdote, grounding conclusions in systematically obtained data rather than assumption or belief.

Measurement, Data Collection, and Analysis

Accurate measurement forms the bridge between observation and interpretation in science. All experiments depend on quantifiable data obtained through calibrated instruments and standardized units. The reliability of findings hinges on precision, accuracy, and consistency in data collection. Scientists employ both direct and indirect measurements, often supplemented by sampling and statistical methods to account for variability. Data analysis transforms raw numbers into meaningful patterns through tools such as graphical representation, descriptive statistics, and inferential testing. Statistical reasoning enables scientists to distinguish real effects from random variation, providing confidence in conclusions. Errors are inevitable but can be minimized through replication and careful methodology. In instruction, teaching data literacy equips students to evaluate scientific claims critically and to recognize bias or misuse of information. Understanding measurement and analysis underscores that science is not defined by certainty but by the disciplined pursuit of increasingly accurate representations of reality.

Use of Models, Simulations, and Technology

Models and simulations serve as conceptual or computational representations of complex systems, enabling scientists to study phenomena that are too vast, small, or abstract for direct observation. Physical models illustrate structural relationships, while mathematical and computer simulations predict behavior under varying conditions. These tools simplify reality while retaining essential variables, allowing exploration of hypotheses and testing of predictions. Technological advances—ranging from digital sensors to satellite imagery—extend human perception, generating vast datasets for analysis. In classrooms, simulations provide interactive opportunities for experimentation without risk or resource limitation. However, models remain approximations; their validity depends on assumptions and parameters. Effective use of modeling teaches students that scientific knowledge evolves through refinement, where simplifications guide understanding until more complete representations emerge. Integrating technology and modeling into instruction fosters both analytical skill and an appreciation of science as an iterative, model-based enterprise.

Laboratory Safety and Ethical Research Practice

The practice of science demands adherence to safety protocols and ethical standards that protect researchers, participants, and the environment. Laboratory safety begins with understanding chemical, biological, and physical hazards and implementing procedures such as proper labeling, waste disposal, and personal protective equipment. Emergency preparedness—knowing the location of fire extinguishers, eyewash stations, and exits—is essential. Equally important are the ethical dimensions of research: honesty in data reporting, respect for living organisms, and compliance with regulatory guidelines. Plagiarism, data fabrication, and manipulation violate the integrity upon which scientific trust depends. Educators play a crucial role in instilling these values, demonstrating that responsible science extends beyond results to the manner in which inquiry is conducted. Safe and ethical practice fosters an environment where discovery proceeds with respect for life, property, and truth, ensuring that scientific progress benefits society without compromising moral or physical safety.

Critical Thinking and Evidence Evaluation

Critical thinking lies at the core of scientific inquiry, enabling the differentiation between evidence-based reasoning and unsupported assertion. It involves questioning assumptions, analyzing sources, recognizing bias, and evaluating the validity of conclusions. Scientists must be able to interpret data objectively, distinguishing correlation from causation and identifying potential sources of error. Peer review institutionalizes critical thinking by subjecting research to scrutiny before acceptance. In education, cultivating this skill transforms students from passive recipients of information into active participants in knowledge construction. By engaging learners in evaluating competing hypotheses or conflicting data, teachers foster intellectual independence and resilience against misinformation. Critical thinking extends beyond science; it equips individuals to assess claims in everyday contexts—health, environment, and technology—where evidence and interpretation intersect. Developing this capacity ensures that inquiry remains both rational and reflective, a discipline of thought as much as a process of discovery.

Communication of Scientific Findings

The dissemination of scientific knowledge transforms individual discovery into collective understanding. Effective communication requires clarity, accuracy, and transparency. Scientists employ multiple formats—

research papers, conference presentations, visual graphics, and digital media—to share methods and results. The structure of scientific writing—introduction, methods, results, discussion—mirrors the logical progression of inquiry, allowing peers to evaluate validity and reproducibility. Visual representation through graphs and diagrams enhances comprehension, while clear language ensures accessibility without compromising precision. Public communication of science bridges the gap between experts and society, fostering trust and informed decision-making. In teaching, writing and presentation assignments cultivate literacy in scientific discourse, reinforcing comprehension through articulation. Communication is not an afterthought to inquiry but an integral phase, where data become knowledge and understanding advances through shared dialogue. The scientific enterprise thus thrives on the open exchange of ideas governed by accuracy, integrity, and respect for evidence.

Interdisciplinary Connections in Science

Modern science transcends traditional boundaries, integrating concepts from physics, chemistry, biology, and earth science to solve complex problems. Interdisciplinary approaches reflect the interconnected nature of reality—climate change involves atmospheric chemistry, ecological biology, and geophysics; medical diagnostics combine genetics, physics, and computer modeling. Collaboration among fields accelerates innovation, fostering new perspectives and technologies. In education, interdisciplinary teaching promotes systems thinking, helping students perceive links between seemingly distinct subjects. This approach mirrors the structure of contemporary research, where advances often occur at the intersection of disciplines. Recognizing these connections deepens comprehension and prepares learners for scientific literacy in an increasingly integrated world. Interdisciplinary inquiry demonstrates that science functions not as isolated silos but as a unified framework for understanding the complexity of nature.

Science Pedagogy and Classroom Application

Effective science instruction translates inquiry into classroom practice, emphasizing exploration, experimentation, and conceptual understanding. Inquiry-based teaching engages students through hands-on investigation, guiding them to construct explanations from evidence rather than memorizing facts. Differentiated instruction accommodates diverse learning styles, using demonstrations, fieldwork, and discussion to reinforce understanding. Technology and digital resources expand access to data and simulations, enhancing engagement and accuracy. The teacher's role evolves from transmitter of knowledge to facilitator of discovery, modeling the habits of mind characteristic of scientists—curiosity, persistence, and reflection. Assessment becomes formative, measuring process as well as product. Through active participation, students internalize not only scientific concepts but the reasoning and ethics that define the discipline. The pedagogical goal is to nurture scientifically literate citizens capable of inquiry and informed judgment in a technologically advanced society.

Assessment, Reflection, and Lifelong Inquiry

Assessment in science education extends beyond grading to encompass reflection and growth. Authentic assessments—laboratory reports, research projects, and presentations—evaluate understanding through application. Reflection encourages students to analyze their thinking processes, recognize misconceptions, and plan further learning. In professional science, peer review and replication serve similar roles, ensuring continuous refinement of knowledge. Lifelong inquiry begins with the recognition that science is never complete; each discovery opens new questions. Teachers foster this perspective by emphasizing curiosity and resilience in the face of uncertainty. The ultimate purpose of science education is not mastery of static facts but the cultivation of an investigative mindset that endures beyond formal schooling. Reflection transforms experience into wisdom, sustaining inquiry as both a professional discipline and a personal way of engaging with the world.

Chapter 9 — Astronomy and the Universe

Ancient Astronomy and the Evolution of Cosmological Models

Astronomy began as humanity's earliest science, emerging from careful observation of the sky to measure time, predict seasons, and guide navigation. Ancient civilizations—Babylonian, Egyptian, Mayan, Chinese, and Greek—recorded celestial motions with remarkable precision. Their observations gave rise to cosmological models that reflected both scientific reasoning and philosophical worldviews. The geocentric model, championed by Ptolemy, placed Earth at the universe's center, with celestial bodies moving in complex epicycles to account for observed motions. Although mathematically sophisticated, this system reflected a belief in the perfection of circular motion and the immutability of the heavens. Over centuries, discrepancies accumulated, prompting the search for alternative explanations. The intellectual evolution from mythological interpretation to geometric modeling marked a pivotal step toward empirical science. Ancient astronomy demonstrated that systematic observation and mathematics could reveal underlying patterns in nature, laying the foundation for modern physics and cosmology. These early inquiries symbolize the enduring human desire to understand order within the apparent vastness of the cosmos.

The Heliocentric Revolution and Modern Planetary Motion

The heliocentric model proposed by Copernicus transformed astronomy by placing the Sun, rather than Earth, at the center of the solar system. This revolutionary idea simplified planetary motion and explained retrograde movement naturally. Galileo's telescopic discoveries—phases of Venus, moons of Jupiter, and sunspots—provided observational evidence supporting heliocentrism and challenged long-held doctrines of celestial perfection. Johannes Kepler refined this model mathematically, formulating three laws of planetary motion: planets orbit the Sun in ellipses, sweep out equal areas in equal times, and maintain a precise relationship between orbital period and distance. Isaac Newton later unified these principles with his law of universal gravitation, showing that the same force governing falling apples holds planets in orbit. Together, these insights replaced philosophical speculation with quantitative prediction, giving rise to celestial mechanics. The heliocentric revolution epitomized the scientific method's power to overturn entrenched ideas through evidence, marking the transition from classical to modern science.

The Solar System: Structure, Formation, and Diversity

The solar system comprises the Sun, eight planets, their moons, and countless smaller bodies—asteroids, comets, and dust. Its formation is explained by the nebular hypothesis, which proposes that a rotating cloud of gas and dust collapsed under gravity approximately 4.6 billion years ago. Conservation of angular momentum caused the cloud to flatten into a disk, with material coalescing into the Sun and planets. Inner planets formed from heavier, refractory elements, while outer planets accumulated lighter gases and ices. This process produced remarkable diversity—from rocky terrestrial worlds to gas and ice giants. Gravitational interactions and residual impacts shaped the system's architecture, creating features such as asteroid belts and orbital resonances. Planetary exploration has revealed ongoing dynamism—volcanism on Io, methane rain on Titan, and subsurface oceans on Europa. Understanding solar system formation illuminates the conditions that make Earth unique and informs the search for habitable worlds beyond our own.

Terrestrial Planets: Geology and Atmospheres

Mercury, Venus, Earth, and Mars share rocky compositions and solid surfaces shaped by volcanism, tectonics, and impact cratering. Mercury's surface records ancient bombardment, while Venus's dense atmosphere generates extreme greenhouse conditions, illustrating how planetary climate can diverge dramatically. Earth remains the only planet with liquid water and an active hydrological cycle, sustained by moderate greenhouse gases and magnetic protection. Mars, once geologically active, retains evidence of ancient river valleys and polar ice caps, suggesting a warmer, wetter past. Comparative planetology reveals how size, composition, and distance from the Sun determine geological evolution and atmospheric retention. The study of terrestrial planets provides context for understanding Earth's long-term stability and vulnerability to change. Their histories demonstrate that planetary environments evolve continuously under the influence of internal heat, solar radiation, and atmospheric chemistry, emphasizing the delicate balance required for habitability within a solar system.

Gas Giants and the Outer Solar System

Beyond the asteroid belt lie the gas giants—Jupiter and Saturn—and the ice giants—Uranus and Neptune. These massive planets dominate the solar system's gravitational structure, influencing the orbits of smaller bodies and protecting inner planets from frequent cometary impacts. Composed primarily of hydrogen, helium, and volatile compounds, they possess deep atmospheres marked by intense weather systems such as Jupiter's Great Red Spot and Saturn's high-latitude storms. Beneath their clouds lie metallic hydrogen layers generating strong magnetic fields and radiating more energy than received from the Sun. Uranus and Neptune, rich in water, ammonia, and methane ices, exhibit distinct bluish hues and dynamic climates driven by internal heat. Their complex satellite systems, including Titan's hydrocarbon lakes and Enceladus's icy plumes, expand our understanding of potential habitability beyond Earth. The gas giants illustrate the diversity of planetary processes and the interplay between chemistry, gravity, and thermodynamics across vast cosmic scales.

Dwarf Planets, Moons, and Small Solar System Bodies

The outer regions of the solar system harbor a vast population of smaller objects—dwarf planets, asteroids, and comets—that preserve primordial material from the system's formation. Pluto, Eris, and Haumea exemplify dwarf planets: icy bodies orbiting beyond Neptune within the Kuiper Belt. Asteroids, primarily located between Mars and Jupiter, vary from metallic fragments to carbon-rich remnants, offering clues to planetary differentiation. Comets, composed of volatile ices and dust, originate in the Kuiper Belt and Oort Cloud, becoming luminous when solar heat vaporizes their surfaces. Moons, diverse in size and geology, often resemble miniature worlds; some, like Europa and Titan, may host subsurface oceans capable of supporting microbial life. Studying these bodies extends planetary science into the frontier of cosmic archaeology, revealing the solar system's early chemistry and dynamic evolution. Their preservation of primitive materials makes them invaluable record keepers of solar history and potential analogs for exoplanetary systems.

The Sun: Structure, Energy, and Solar Activity

The Sun, a G-type main-sequence star, anchors the solar system through gravity and radiation. Its energy arises from nuclear fusion in the core, where hydrogen nuclei combine to form helium, releasing immense energy according to Einstein's equation, $E = mc^2$. This energy travels outward through the radiative and convective zones before escaping as sunlight. The photosphere, visible surface of the Sun, displays granulation and sunspots—manifestations of magnetic activity. The overlying chromosphere and corona extend millions of kilometers, generating solar flares and coronal mass ejections that influence Earth's magnetosphere and climate. The solar cycle, approximately eleven years, reflects periodic magnetic reversals. Understanding solar processes is critical not only for astronomy but also for protecting technology-dependent societies from space weather effects. The Sun exemplifies stellar stability and energy transformation, demonstrating how fusion sustains life on Earth while linking planetary systems to stellar evolution.

Stars, Galaxies, and the Scale of the Universe

Stars form within molecular clouds, where gravity compresses gas and dust until nuclear fusion ignites. Stellar properties—mass, temperature, luminosity—determine their life cycles. Massive stars burn rapidly, ending as supernovae or black holes; smaller stars evolve more slowly, forming white dwarfs. Groups of stars bound by gravity form galaxies, which exist in diverse forms—spiral, elliptical, and irregular. The Milky Way, our galaxy, spans over 100,000 light-years and contains billions of stars organized around a supermassive black hole. Clusters of galaxies assemble into vast superclusters, defining the large-scale structure of the universe. Observations of light across the electromagnetic spectrum reveal that the universe extends far beyond human perception, governed by universal physical laws. The study of stars and galaxies transforms human understanding of scale and time, revealing that cosmic processes mirror the same physics observed in laboratories, uniting microcosm and macrocosm under one framework of natural order.

Cosmology: The Big Bang, Expansion, and Dark Matter

Cosmology seeks to understand the origin, structure, and fate of the universe. The prevailing model, the Big Bang theory, proposes that the universe began as a hot, dense state approximately 13.8 billion years ago and has been expanding ever since. Evidence includes cosmic microwave background radiation, the abundance of light elements, and the redshift of distant galaxies. Observations indicate that the expansion is accelerating, driven by an unknown component called dark energy, while dark matter—an invisible form of mass—accounts for much of the universe's gravitational structure. Together, these unseen forces shape galaxies and cosmic evolution. Modern cosmology combines observation with theoretical physics to probe fundamental questions about space, time, and matter. The study of cosmology exemplifies the intellectual ambition of science: to trace the universe from its first moments to its potential future, revealing that the cosmos is both ancient and continuously unfolding.

Observation, Exploration, and Humanity's Place in the Cosmos

Advances in telescopes and spacecraft have transformed astronomy from remote observation to direct exploration. From Galileo's refractor to modern observatories orbiting Earth and operating on distant planets, technology extends human vision across space and time. Robotic missions map planetary surfaces, analyze atmospheres, and search for biosignatures beyond Earth. The discovery of thousands of exoplanets demonstrates that planetary systems are common throughout the galaxy, renewing questions about life elsewhere. Human spaceflight, from the Moon to the International Space Station, symbolizes both technological achievement and philosophical reflection. Astronomy ultimately connects physical observation with existential inquiry—it reveals humanity's smallness yet affirms our capacity to comprehend the vast. The pursuit of cosmic knowledge transcends utility; it embodies the enduring human aspiration to explore, understand, and find meaning within the infinite expanse of the universe.

Chapter 10 — Science, Technology, and Society

Interdependence of Science and Technology

Science and technology exist in a continuous cycle of mutual reinforcement. Scientific discovery expands theoretical understanding, while technology applies that knowledge to practical use, often revealing new phenomena in return. For example, advances in electromagnetism led to electric power generation, which in turn enabled new experimental methods in physics and chemistry. Likewise, digital technology now facilitates data collection and analysis, accelerating every field of research. This symbiosis illustrates that neither science nor technology operates in isolation; both evolve through shared curiosity and problem-solving. The iterative relationship between theory and application transforms society by redefining how humans interact with the world. Recognizing this interdependence helps students understand that innovation is not merely invention but the systematic application of scientific reasoning to human needs. In education and policy, integrating scientific literacy with technological competence ensures that progress remains informed, ethical, and sustainable, linking knowledge creation with responsibility to both people and planet.

Historical Transformations and the Scientific Revolution

The Scientific Revolution of the seventeenth century redefined humanity's understanding of nature and knowledge. By emphasizing observation, experimentation, and mathematical reasoning, thinkers such as Galileo, Kepler, and Newton established a method of inquiry that replaced authority with evidence. This intellectual shift laid the groundwork for industrial and technological revolutions, transforming economies and societies. The scientific worldview—rooted in empiricism and testable hypotheses—spread beyond laboratories into agriculture, medicine, and governance. The Enlightenment extended these ideas, linking scientific progress with human rights and rational education. Every subsequent era of innovation, from the mechanization of industry to the digital age, traces its lineage to this transformation. Understanding the history of science reveals that progress arises not from isolated genius but from cumulative knowledge built upon skepticism, collaboration, and open exchange. For educators, this legacy underscores that scientific thinking is both a method of discovery and a moral framework guiding the responsible use of knowledge.

Industrialization, Innovation, and Environmental Change

Industrialization marked a turning point in human history, multiplying production and connectivity while reshaping the environment. The mechanization of labor through steam power and electricity revolutionized transportation, manufacturing, and agriculture. Yet these achievements also introduced new challenges—resource depletion, pollution, and social inequality. The same technologies that enhanced comfort and efficiency also disrupted ecosystems and altered climate through massive carbon emissions. The modern environmental movement emerged as a response to these unintended consequences, calling for technologies that balance growth with sustainability. The relationship between industry and environment illustrates the double-edged nature of progress: human ingenuity can both create and endanger. Understanding this duality enables societies to evaluate innovation not only for its benefits but also for its long-term costs. In teaching science and technology, highlighting this balance fosters critical awareness of humanity's capacity to shape—and safeguard—the biosphere upon which civilization depends.

Information Technology and the Digital Revolution

The digital revolution redefined communication, knowledge, and social interaction. The development of microprocessors, the Internet, and global networks transformed the speed and scale at which information is created and shared. Data now drive decision-making in science, economics, and governance, creating opportunities for collaboration but also raising questions about privacy, accuracy, and equity. Artificial intelligence and machine learning extend human analytical power while challenging traditional notions of authorship and responsibility. In education, digital platforms democratize access to information, yet they also demand new forms of literacy to navigate misinformation and digital manipulation. The integration of computing with biology, astronomy, and environmental modeling illustrates how technology magnifies scientific potential. However, it also underscores the ethical obligation to ensure that data serve humanity rather than exploit it. Understanding the digital revolution requires balancing technological optimism with critical awareness, ensuring that innovation remains guided by wisdom as well as speed.

Medicine, Biotechnology, and Human Health

Scientific progress in medicine has extended human life and transformed quality of health. From the discovery of germ theory and antibiotics to the decoding of the human genome, biology and technology now intersect at unprecedented depth. Biotechnology enables targeted therapies, genetic engineering, and regenerative medicine, offering solutions to diseases once thought incurable. Yet these advances raise profound ethical questions concerning privacy, equity, and the definition of human enhancement. Vaccination, organ transplantation, and reproductive technology illustrate how science can both heal and provoke debate about moral boundaries. Public health depends not only on innovation but also on education and equitable access. Understanding medical technology as part of the broader fabric of society emphasizes that health is both a biological and social condition. The study of biotechnology in education fosters scientific literacy grounded in compassion, preparing future citizens to evaluate the promises and limits of biomedical progress responsibly.

Energy, Industry, and Global Sustainability

Energy underpins all technological development, connecting physical science with economic and environmental systems. The transition from traditional fuels to renewable sources defines one of the central challenges of modern civilization. Fossil fuels enabled industrial growth but contributed to greenhouse gas accumulation and climate instability. Sustainable energy—solar, wind, geothermal, and tidal—offers alternatives that align human progress with ecological limits. Energy policy thus becomes both a scientific and ethical issue, balancing accessibility, affordability, and environmental integrity. Innovation in storage, grid technology, and efficiency demonstrates how applied physics and engineering can address global needs. Understanding energy systems requires integrating thermodynamics, economics, and ecology into a unified framework of sustainability. Educating about energy encourages awareness of the consequences of consumption, highlighting that responsible technology is not defined by what is possible but by what is sustainable for generations to come.

Globalization, Equity, and Technological Access

Technology has interconnected the world, accelerating communication, commerce, and culture. However, globalization also exposes disparities in access to knowledge and resources. The digital divide separates communities with technological infrastructure from those without, perpetuating inequality in education, healthcare, and opportunity. Scientific advancement benefits humanity only when its fruits are equitably distributed. Addressing inequity involves international cooperation, capacity building, and inclusive innovation that respects cultural and environmental diversity. Technology must serve as a tool for empowerment rather than exploitation. In the classroom, discussing global access connects ethics with science policy, reminding students that discovery carries social responsibility. Sustainable globalization integrates technology with justice, ensuring that progress uplifts rather than divides. The vision of a connected world thus depends not solely on circuits and algorithms but on empathy, fairness, and shared stewardship of human knowledge.

Science, Ethics, and Responsibility

Every scientific and technological advance carries ethical dimensions. Questions of safety, consent, environmental impact, and social justice accompany research in genetics, artificial intelligence, and energy production. Ethics ensures that innovation serves humanity rather than undermines it. Scientists must weigh the potential consequences of their work, guided by principles of integrity, transparency, and respect for life. Historical examples—from nuclear technology to environmental contamination—illustrate the consequences of neglecting ethical reflection. Modern science incorporates institutional safeguards such as review boards and international codes, yet individual conscience remains central. Teaching ethics in science education cultivates moral reasoning alongside technical knowledge, preparing students to act as responsible innovators. Ultimately, ethics transforms knowledge into wisdom, aligning discovery with the values of compassion and stewardship that sustain civilization's trust in science.

Education, Literacy, and the Public Understanding of Science

Scientific literacy empowers citizens to participate meaningfully in democratic societies shaped by technology. Public understanding of science determines how policies on health, environment, and innovation are formed and accepted. Education must therefore cultivate not only factual knowledge but also curiosity, skepticism, and critical reasoning. Communication between scientists and the public bridges cultural gaps and prevents the spread of misinformation. Media, museums, and digital platforms serve as conduits for engagement, translating complex ideas into accessible narratives. Teachers play a pivotal role by modeling inquiry and emphasizing that uncertainty is a natural part of scientific progress. A scientifically literate population can evaluate evidence, recognize bias, and make informed choices about technology and policy. In this way, education becomes the foundation of an enlightened society capable of guiding innovation toward shared benefit rather than division.

The Future of Science and Human Civilization

The future of science lies in integration—uniting disciplines, cultures, and perspectives to address planetary challenges. Artificial intelligence, space exploration, and genetic engineering promise profound transformation, but they also demand humility in the face of complexity. The sustainability of civilization will depend on whether technology advances faster than wisdom or in harmony with it. Future scientists must balance exploration with ethics, ambition with stewardship. As humanity extends its reach beyond Earth and into digital frontiers, the guiding principle remains constant: knowledge must serve life. Education, policy, and collaboration will determine whether science becomes an instrument of preservation or disruption. The ultimate goal is not control over nature but coexistence within it—an enduring partnership between intelligence and environment. Science, in its highest form, expresses humanity's capacity to understand creation and, through that understanding, to care for it wisely.

Chapter 11 – Practice Questions

Questions (1-100)

Q1. Which of the following best defines the scientific method?
A. A set of beliefs supported by personal opinion
B. A structured process of observation, hypothesis, experimentation, and conclusion
C. Random trial and error guided by chance
D. A collection of unverified assumptions

Q2. Which principle ensures that scientific knowledge remains open to revision?
A. Authority
B. Tradition
C. Falsifiability
D. Confirmation bias

Q3. A hypothesis must be:
A. Testable and falsifiable
B. Unverifiable speculation
C. A statement of belief
D. Proven fact

Q4. Why is peer review essential in scientific inquiry?
A. It guarantees perfect results
B. It prevents publication delays
C. It ensures scientific validity through critical evaluation
D. It simplifies the experimental design

Q5. Which of the following best illustrates deductive reasoning in science?
A. Developing a theory after observing patterns
B. Testing a general law to predict specific outcomes
C. Guessing experimental results without data
D. Collecting random data for analysis

Q6. Inductive reasoning in science involves:
A. Moving from general laws to specific observations
B. Applying untested assumptions
C. Drawing broad conclusions from specific observations
D. Ignoring experimental data

Q7. Which of the following ensures objectivity in experimentation?
A. Confirmation bias
B. Use of control groups and replication
C. Ignoring anomalous results
D. Relying solely on personal intuition

Q8. Which of the following best represents a quantitative observation?
A. The chemical smelled strong
B. The reaction appeared faster
C. The color turned darker
D. The temperature increased by 5°C

Q9. In experimental design, the independent variable is:
A. The factor being measured
B. The factor deliberately manipulated
C. The outcome of the experiment
D. The uncontrolled environmental factor

Q10. The dependent variable in an experiment:
A. Is kept constant
B. Represents the manipulated factor
C. Shows the measurable response
D. Is irrelevant to hypothesis testing
Q11. The control group in a scientific experiment serves to:
A. Receive identical treatment as the experimental group
B. Provide a baseline for comparison
C. Test multiple hypotheses at once
D. Eliminate the need for repetition
Q12. Why are repeated trials important in scientific experiments?
A. They eliminate all errors permanently
B. They provide statistical reliability
C. They simplify the experimental process
D. They guarantee positive results
Q13. Which of the following demonstrates ethical scientific conduct?
A. Manipulating data to fit hypotheses
B. Reporting all results, even unexpected ones
C. Withholding methods to protect competition
D. Ignoring contradictory evidence
Q14. Which of the following best describes a scientific theory?
A. An untested guess
B. A provisional idea with no evidence
C. A well-substantiated explanation supported by evidence
D. A simple hypothesis
Q15. Which statement reflects the relationship between theories and laws?
A. Theories become laws after proof
B. Laws explain theories
C. Theories explain laws
D. They are identical concepts
Q16. Which type of scientific investigation compares groups without manipulation?
A. Experimental
B. Observational
C. Quasi-experimental
D. Simulation
Q17. In the context of the scientific method, what follows after forming a hypothesis?
A. Drawing conclusions immediately
B. Revising laws of science
C. Publishing results before testing
D. Conducting controlled experimentation
Q18. The reproducibility of results primarily supports which aspect of science?
A. Creativity
B. Objectivity
C. Individual interpretation
D. Ethical subjectivity
Q19. Which concept emphasizes the use of prior research to guide new investigations?
A. Literature review
B. Replication bias

C. Random sampling
D. Controlled variation

Q20. What distinguishes scientific explanations from non-scientific ones?
A. Personal belief
B. Empirical evidence and testability
C. Authority endorsement
D. Cultural tradition

Q21. The principle of parsimony (Occam's Razor) advises scientists to:
A. Prefer complex explanations
B. Accept untestable hypotheses
C. Choose the simplest explanation that fits the evidence
D. Ignore conflicting data

Q22. Why is operational definition critical in scientific inquiry?
A. It ensures measurable and replicable variables
B. It restricts experimental freedom
C. It prevents statistical analysis
D. It simplifies complex theories

Q23. Which of the following best describes an experimental constant?
A. The factor deliberately changed
B. The measured outcome
C. The element kept unchanged
D. The unexpected result

Q24. A scientist designs an experiment to test fertilizer effects on plant growth. Which is the dependent variable?
A. Amount of fertilizer applied
B. Growth rate of plants
C. Type of soil used
D. Type of fertilizer

Q25. What is the role of modeling in scientific investigation?
A. To replace experiments entirely
B. To restrict creativity in inquiry
C. To eliminate empirical data collection
D. To provide visual and conceptual representations of complex systems

Q26. Why is replication by independent researchers crucial in science?
A. It confirms findings and minimizes bias
B. It accelerates publication timelines
C. It allows subjective interpretation
D. It eliminates the need for peer review

Q27. Which best characterizes empirical evidence?
A. Derived from logical reasoning alone
B. Based on observable and measurable data
C. Formed through anecdotal reports
D. Determined by intuition or faith

Q28. The reliability of an experiment refers to:
A. The experiment's novelty
B. Accuracy of a single measurement
C. Consistency of results across repeated trials
D. Theoretical alignment with established laws

Q29. Which statement accurately describes validity in experimentation?
A. It measures whether results are consistent
B. It indicates whether the experiment tests what it claims to test
C. It ensures data collection is fast
D. It guarantees identical outcomes every time
Q30. Why is it important for scientific data to be transparent and publicly accessible?
A. To allow for verification and reproducibility
B. To protect researcher confidentiality
C. To prevent peer criticism
D. To reduce workload in analysis
Q31. What distinguishes an experimental hypothesis from a null hypothesis?
A. The null hypothesis asserts no effect, while the experimental predicts an effect
B. The experimental hypothesis denies all variables
C. The null hypothesis confirms experimental bias
D. They represent identical predictions
Q32. Which of the following represents a qualitative observation?
A. The solution's pH is 6.4
B. The mass of the sample is 25 g
C. The plant leaves turned yellow
D. The temperature rose 2°C
Q33. Why do scientists use statistical analysis?
A. To manipulate data subjectively
B. To interpret data objectively and test significance
C. To avoid data collection
D. To reduce sample size
Q34. The main advantage of double-blind experimental design is:
A. It eliminates all random error
B. It reduces bias from both participants and researchers
C. It guarantees positive results
D. It simplifies statistical analysis
Q35. Why is measurement precision important in scientific research?
A. It ensures random variation
B. It replaces statistical analysis
C. It decreases equipment costs
D. It increases accuracy and reproducibility of data
Q36. In scientific reasoning, what does a model's predictive power indicate?
A. Its simplicity
B. Its ability to forecast outcomes accurately
C. Its historical popularity
D. Its reliance on subjective judgment
Q37. The term "control variable" refers to:
A. The variable being measured
B. The variable intentionally manipulated
C. A factor kept constant to prevent confounding
D. A random error in the system
Q38. Which process converts raw data into meaningful scientific conclusions?
A. Hypothesis elimination
B. Data interpretation and analysis

C. Random sampling
D. Uncontrolled observation

Q39. Why are negative results valuable in science?
A. They indicate experimental failure
B. They refine hypotheses and prevent bias
C. They reduce publication quality
D. They invalidate the scientific method

Q40. Which factor strengthens external validity in experiments?
A. Large, diverse sample sizes
B. Small, homogeneous groups
C. Arbitrary selection of data
D. Ignoring replication

Q41. Which of the following illustrates bias in scientific research?
A. Publishing full datasets
B. Selecting data that supports a hypothesis
C. Peer collaboration
D. Randomized sampling

Q42. Why is reproducibility a key characteristic of scientific knowledge?
A. It proves the superiority of a theory
B. It ensures results can be independently verified
C. It eliminates uncertainty entirely
D. It limits creative exploration

Q43. Which of the following represents a testable hypothesis?
A. Love improves health
B. Music makes people happy
C. Meditation reduces heart rate after 10 minutes
D. Anger causes misfortune

Q44. Which best describes the role of skepticism in science?
A. Rejecting all new ideas
B. Critically evaluating evidence before acceptance
C. Accepting authority unquestioningly
D. Avoiding peer collaboration

Q45. What is the purpose of using a placebo in controlled trials?
A. To increase sample variability
B. To measure psychological effects separate from treatment
C. To speed up analysis
D. To prove the treatment ineffective

Q46. What is the main difference between accuracy and precision?
A. Accuracy refers to closeness to true value; precision reflects repeatability
B. Accuracy indicates reproducibility
C. Precision means correctness
D. They are identical concepts

Q47. Why do scientists use control conditions in experimental design?
A. To manipulate all variables
B. To compare changes due to the independent variable
C. To ensure identical outcomes
D. To limit sample size

Q48. Which of the following reflects pseudoscience rather than science?

A. Claims not open to testing or falsification
B. Data gathered through experimentation
C. Hypotheses verified by replication
D. Transparent peer review

Q49. Which feature most distinguishes science from other forms of knowledge?
A. Its dependence on authority
B. Its empirical and testable nature
C. Its focus on tradition
D. Its emotional influence

Q50. Why is clear communication of results critical to scientific progress?
A. It allows collaboration, verification, and further research
B. It restricts peer involvement
C. It conceals methodology
D. It minimizes replication

Q51. Which step in the scientific method involves comparing results to the hypothesis?
A. Observation
B. Experimentation
C. Hypothesis formulation
D. Data analysis and conclusion

Q52. What is the significance of using standardized measurement units in science?
A. They ensure global consistency and comparability of data
B. They limit research flexibility
C. They prevent replication
D. They promote subjective interpretation

Q53. Which statement best defines a variable?
A. A fixed condition in an experiment
B. Any factor that can change or be controlled
C. A predetermined conclusion
D. A random event

Q54. Why are models revised in scientific research?
A. To reflect new evidence or improved understanding
B. To discard scientific methods
C. To ensure secrecy of results
D. To fit predetermined conclusions

Q55. What role does creativity play in scientific inquiry?
A. It hinders logical reasoning
B. It drives hypothesis formation and problem-solving
C. It replaces empirical validation
D. It relies on superstition

Q56. Which type of research aims primarily at increasing basic knowledge rather than immediate application?
A. Applied research
B. Theoretical research
C. Basic research
D. Operational research

Q57. Which of the following is a limitation of the scientific method?
A. It cannot answer moral or aesthetic questions
B. It fails to produce consistent evidence

C. It eliminates uncertainty entirely
D. It depends solely on authority

Q58. Which term refers to the process of collecting and analyzing previous studies before starting new research?

A. Experimental analysis
B. Theoretical modeling
C. Literature review
D. Qualitative synthesis

Q59. Why is the reproducibility crisis a major concern in modern science?

A. It questions the reliability of published results
B. It increases research funding
C. It validates all studies automatically
D. It simplifies peer review

Q60. Which of the following is an example of applied research?

A. Investigating genetic code structure
B. Developing a vaccine for a new virus
C. Studying atomic theory development
D. Exploring planetary motion

Q61. Why do scientists use blind procedures in experiments?

A. To reduce observer or participant bias
B. To hide unethical practices
C. To make analysis easier
D. To eliminate the need for controls

Q62. What is the importance of uncertainty in scientific measurement?

A. It represents measurement error and defines confidence limits
B. It invalidates scientific data
C. It prevents result interpretation
D. It proves all measurements wrong

Q63. What is a paradigm shift in science?

A. Gradual improvement of theories
B. Sudden replacement of one scientific framework by another
C. Minor correction of experimental errors
D. Temporary research trend

Q64. Which factor most enhances the credibility of scientific data?

A. Peer review and independent verification
B. Limited sample selection
C. Anecdotal reporting
D. Rapid publication

Q65. What differentiates correlation from causation?

A. Correlation proves direct cause
B. Correlation indicates association, not causality
C. Causation depends on random events
D. They are equivalent concepts

Q66. Which principle of scientific ethics demands that data not be fabricated or falsified?

A. Transparency
B. Confidentiality
C. Replicability
D. Integrity

Q67. Which method ensures fairness in assigning subjects to experimental and control groups?
A. Randomization
B. Selection bias
C. Stratified grouping
D. Controlled error
Q68. Why do scientists formulate null hypotheses in research?
A. To simplify data interpretation
B. To provide a basis for statistical testing
C. To ensure results are positive
D. To avoid peer review
Q69. What is the role of replication studies in scientific progress?
A. They confirm or challenge previous findings
B. They prevent innovation
C. They duplicate errors intentionally
D. They reduce public trust
Q70. Why is calibration important in using scientific instruments?
A. It adjusts tools to known standards for accuracy
B. It reduces observation frequency
C. It simplifies statistical work
D. It eliminates all uncertainty
Q71. Which type of reasoning combines both induction and deduction in scientific thinking?
A. Analogical reasoning
B. Abductive reasoning
C. Circular reasoning
D. Descriptive reasoning
Q72. Why do scientists publish detailed methodologies with their results?
A. To increase complexity unnecessarily
B. To obscure analytical techniques
C. To enable replication and verification
D. To limit peer collaboration
Q73. Which feature of scientific inquiry promotes continuous improvement of knowledge?
A. Static theories
B. Open-mindedness to new evidence
C. Reliance on tradition
D. Rejection of peer critique
Q74. What is the primary goal of a pilot study?
A. To test feasibility and refine methods before full-scale research
B. To prove hypotheses conclusively
C. To avoid statistical analysis
D. To finalize conclusions
Q75. Which characteristic best defines the nature of science?
A. Static knowledge based on authority
B. Dynamic, evidence-based process open to revision
C. Collection of unverifiable opinions
D. Immutable system of facts
Q76. Which of the following quantities is a vector?
A. Speed
B. Distance

C. Work
D. Velocity
E. Power

Q77. A car moves in a circular path at a constant speed. Which of the following statements is true?

A. The car has no acceleration.
B. The car's acceleration is directed outward.
C. The car's acceleration is directed toward the center.
D. The car's velocity remains constant.
E. The car's kinetic energy decreases.

Q78. A 2-kg object experiences a net force of 10 N. What is its acceleration?

A. 0.2 m/s^2
B. 2 m/s^2
C. 5 m/s^2
D. 10 m/s^2
E. 20 m/s^2

Q79. Which of the following is an example of Newton's Third Law?

A. A rocket accelerates upward while expelling gas downward.
B. A ball rolling to a stop due to friction.
C. A falling apple accelerating toward Earth.
D. A car speeding up from rest.
E. A pendulum changing direction at its peak.

Q80. The momentum of an object is defined as:

A. Force divided by time
B. Mass times acceleration
C. Mass times velocity
D. Force times distance
E. Energy divided by time

Q81. What happens to kinetic energy when the velocity of an object doubles?

A. It remains the same.
B. It doubles.
C. It triples.
D. It quadruples.
E. It increases eightfold.

Q82. Which of the following best describes work done on an object?

A. The change in its velocity
B. The product of force and distance in the direction of force
C. The area under a velocity-time graph
D. The energy radiated by heat
E. The inverse of momentum

Q83. In an elastic collision between two objects:

A. Momentum is conserved but kinetic energy is not.
B. Kinetic energy is conserved but momentum is not.
C. Both momentum and kinetic energy are conserved.
D. Neither is conserved.
E. Only angular momentum is conserved.

Q84. What physical quantity is represented by the slope of a velocity-time graph?

A. Displacement
B. Distance

C. Acceleration
D. Momentum
E. Power

Q85. The first law of thermodynamics states:
A. Energy cannot be created or destroyed, only transformed.
B. The entropy of a system always decreases.
C. Heat flows spontaneously from cold to hot.
D. Pressure is inversely proportional to volume.
E. Mass increases with velocity.

Q86. The SI unit of power is:
A. Joule
B. Watt
C. Newton
D. Pascal
E. Ampere

Q87. Which of the following best explains why satellites stay in orbit?
A. They have no gravity acting on them.
B. Their engines continually fire upward.
C. Their forward velocity balances gravitational pull.
D. They are outside Earth's gravitational field.
E. Centrifugal force repels them.

Q88. Which statement about electromagnetic waves is correct?
A. They require a medium to travel.
B. Their speed depends on frequency.
C. They are longitudinal waves.
D. They consist of oscillating electric and magnetic fields.
E. They transport matter.

Q89. A wave's energy is proportional to:
A. Its wavelength
B. Its frequency squared
C. Its amplitude squared
D. Its period
E. Its phase angle

Q90. According to Ohm's Law, current in a resistor:
A. Increases with voltage and resistance
B. Increases with voltage and decreases with resistance
C. Decreases with both voltage and resistance
D. Remains constant for all voltages
E. Is independent of resistance

Q91. A simple pendulum converts energy between which forms during its motion?
A. Chemical and electrical
B. Thermal and nuclear
C. Kinetic and potential
D. Magnetic and electric
E. Gravitational and elastic

Q92. In projectile motion (neglecting air resistance), which statement is true about horizontal and vertical components?
A. Both components change due to gravity

B. Horizontal velocity changes, vertical is constant
C. Horizontal is constant, vertical changes due to gravity
D. Both components are constant
E. Vertical velocity changes sign but magnitude remains constant

Q93. The centripetal force required for circular motion of an object of mass m and speed v on radius r is:

A. mv/r
B. mvr
C. mv^2/r
D. mr^2/v
E. v^2/(mr)

Q94. After an explosion breaks a stationary object into fragments with no external forces, which is conserved?

A. Kinetic energy only
B. Momentum only
C. Momentum and kinetic energy
D. Neither momentum nor kinetic energy
E. Angular momentum only

Q95. Hydrostatic pressure at depth h in a fluid of density ρ is given by:

A. p = ρgh
B. p = ρg/h
C. p = gh/ρ
D. p = ρg + h
E. p = g/ρh

Q96. The Doppler effect causes observed frequency to change when:

A. The source changes temperature
B. The observer or source moves relative to the medium
C. Amplitude of the wave changes
D. Wavelength becomes constant
E. The medium becomes denser

Q97. Photon energy is related to frequency by which expression?

A. E = mc^2
B. E = 1/2 mv^2
C. E = hf
D. E = kT
E. E = qV

Q98. The time constant τ for an RC circuit (resistor R and capacitor C) determines:

A. The steady-state current only
B. The initial voltage only
C. The characteristic charging/discharging time scale
D. The power dissipated in R
E. The resonance frequency

Q99. Diffraction limits resolution by causing waves to spread when passing through apertures. The smallest resolvable angle θ for a circular aperture is approximately:

A. θ ≈ D/λ
B. θ ≈ λ/D
C. θ ≈ λD
D. θ ≈ 1/(λD)

E. $\theta \approx (\lambda+D)/2$

Q100. Bernoulli's principle contributes to lift on an airfoil because:

A. Higher pressure above the wing creates lift

B. Faster flow above the wing lowers pressure above, aiding lift

C. Wings are glued to the air with suction

D. Lift is caused solely by wing weight

E. Bernoulli's principle does not apply to airfoils

Answer keys and Explanations (1-100)

Q1. Answer: B

Explanation: The scientific method is a systematic approach involving observation, hypothesis formulation, controlled experimentation, and evidence-based conclusions. It provides a repeatable, objective means of establishing causal relationships. Unlike random guessing, it depends on verifiable and falsifiable testing, ensuring reproducibility and critical evaluation of findings to advance knowledge through rigorous empirical validation.

Q2. Answer: C

Explanation: Falsifiability allows scientific claims to be tested and potentially disproven, ensuring science evolves with new evidence. It distinguishes science from dogma by requiring hypotheses to be testable and modifiable through data. This dynamic nature enables theories to adapt to new discoveries, maintaining the integrity of science as a self-correcting and progressive system of inquiry.

Q3. Answer: A

Explanation: A hypothesis is a proposed explanation that must be testable and falsifiable through empirical evidence. It functions as a predictive statement guiding experimentation. Unlike opinions or facts, hypotheses are tentative, subject to testing and potential rejection. Their scientific validity depends on measurable outcomes and the capacity for replication by independent researchers.

Q4. Answer: C

Explanation: Peer review subjects research to scrutiny by experts, verifying methodology, data interpretation, and logical reasoning. This process upholds scientific quality, detects flaws or biases, and promotes transparency. Although it cannot guarantee perfection, it fosters reliability and collective accountability, maintaining confidence in published findings across the scientific community.

Q5. Answer: B

Explanation: Deductive reasoning applies general principles or laws to predict specific outcomes, moving from the general to the particular. It allows scientists to derive testable predictions based on established theories. This logical approach ensures coherence between theory and observation, distinguishing structured scientific reasoning from mere speculation or arbitrary data collection.

Q6. Answer: C

Explanation: Inductive reasoning generalizes from particular observations to form broader principles or theories. It underpins hypothesis generation and model development, enabling scientists to infer consistent patterns. Though inductive conclusions are probabilistic, their strength lies in repeated, reliable observations that gradually build coherent scientific laws grounded in empirical evidence.

Q7. Answer: B

Explanation: Objectivity in science is maintained through controlled conditions, replication, and elimination of bias. Control groups provide baselines for comparison, while repeated trials confirm consistency. By minimizing subjective influence, these practices ensure that observed outcomes reflect genuine causal relationships, reinforcing scientific reliability and reproducibility.

Q8. Answer: D

Explanation: Quantitative observations involve measurable, numerical data such as temperature, volume, or mass. They provide precision and allow for statistical analysis, unlike qualitative descriptions, which rely on subjective interpretation. Quantitative data thus strengthen the accuracy and comparability of scientific findings, forming the backbone of empirical validation.

Q9. Answer: B

Explanation: The independent variable is intentionally altered to observe its effect on the dependent variable. It allows causal relationships to be established by systematic manipulation. Proper control of other variables ensures that changes in outcomes result solely from the independent variable, maintaining the experiment's internal validity and interpretability.

Q10. Answer: C

Explanation: The dependent variable measures the effect of changes in the independent variable, serving as the experiment's outcome indicator. It reflects the hypothesized relationship under investigation. Accurate measurement and control of confounding variables ensure that observed differences are meaningful and statistically significant within the tested framework.

Q11. Answer: B

Explanation: The control group acts as a reference, remaining unaltered by the independent variable to reveal genuine treatment effects. It isolates the influence of manipulated factors, ensuring observed outcomes are not due to external variables. This comparative structure underpins the validity of experimental conclusions in scientific investigations.

Q12. Answer: B

Explanation: Repetition increases confidence in data by reducing random error and verifying consistency across trials. Statistical reliability arises when results are reproducible under similar conditions. This practice distinguishes robust scientific conclusions from coincidental findings, ensuring that patterns observed represent genuine causal relationships rather than anomalies.

Q13. Answer: B

Explanation: Ethical conduct in science requires transparency and honesty in data reporting, including anomalies. Publishing complete, accurate findings fosters trust and enables replication. Concealing or altering results violates the principles of integrity and objectivity, undermining the self-correcting nature and credibility of the scientific enterprise.

Q14. Answer: C

Explanation: A scientific theory synthesizes validated hypotheses into a coherent explanatory framework supported by extensive empirical evidence. It predicts phenomena and guides further investigation. Unlike guesses or single hypotheses, theories represent tested, reliable understandings continually refined as new observations emerge, forming the cornerstone of scientific progress.

Q15. Answer: C

Explanation: Laws describe consistent natural patterns through concise mathematical or verbal statements, while theories explain the mechanisms behind those patterns. Theories never become laws; they coexist, serving distinct roles. Theories interpret why phenomena occur, whereas laws describe how they occur, collectively enhancing predictive scientific understanding.

Q16. Answer: B

Explanation: Observational studies record natural phenomena without altering variables, identifying correlations and patterns. They are vital in fields where controlled experiments are impractical or unethical, such as ecology or astronomy. While they cannot establish causality, they generate valuable data for hypothesis formation and future experimental testing.

Q17. Answer: D

Explanation: After hypothesis formation, researchers design experiments to test predicted relationships systematically. This step validates or refutes hypotheses through empirical observation. Experimental evidence provides the foundation for logical conclusions, ensuring that inferences arise from data rather than assumption or speculation.

Q18. Answer: B

Explanation: Reproducibility ensures objectivity by demonstrating that results are consistent across independent trials and researchers. It confirms findings are not artifacts of personal bias or random error. This standard enables cumulative knowledge building, allowing the scientific community to validate discoveries through transparent, repeatable methodologies.

Q19. Answer: A

Explanation: A literature review synthesizes prior findings, identifying knowledge gaps and refining research questions. It contextualizes new studies within existing frameworks, preventing redundancy and supporting informed hypothesis formulation. This process anchors new inquiries in the broader scientific dialogue, ensuring coherence and continuity in discovery.

Q20. Answer: B

Explanation: Scientific explanations rely on empirical evidence and testability, separating them from personal or cultural beliefs. Their validity arises from systematic observation, experimentation, and verification, enabling replication and falsification. Non-scientific explanations lack these criteria, preventing objective assessment or predictive capability within the scientific framework.

Q21. Answer: C

Explanation: Occam's Razor promotes simplicity by favoring explanations requiring the fewest assumptions while accounting for observed data. It prevents overcomplication and helps maintain conceptual clarity. Simplicity enhances testability and replicability, ensuring that scientific models remain efficient, coherent, and grounded in observable evidence rather than speculation.

Q22. Answer: A

Explanation: Operational definitions specify how variables are measured or manipulated, ensuring consistency and reproducibility. By clearly defining terms and procedures, they eliminate ambiguity and allow other researchers to verify findings. This precision strengthens validity, enabling comparison and replication across studies within the scientific community.

Q23. Answer: C

Explanation: Constants are controlled elements maintained throughout experimentation to isolate variable effects. They safeguard validity by ensuring observed changes arise solely from the independent variable. Maintaining constants enables reliable interpretation of data and enhances the accuracy of cause-and-effect relationships derived from experimental designs.

Q24. Answer: B

Explanation: The dependent variable, plant growth rate, reflects the response measured to assess the fertilizer's impact. Independent variables, such as fertilizer amount, are manipulated to determine causality. Accurate measurement of dependent outcomes is essential for evaluating hypotheses and ensuring empirical credibility in agricultural experimentation.

Q25. Answer: D

Explanation: Models simplify and represent intricate systems, allowing scientists to visualize, predict, and test theoretical relationships. They bridge abstract concepts and empirical data, facilitating understanding and communication. Models evolve with evidence, reinforcing their role as essential tools for simulation, hypothesis refinement, and explanatory development in scientific inquiry.

Q26. Answer: A

Explanation: Replication by independent researchers validates findings, minimizing individual bias and error. It demonstrates that results are consistent, reproducible, and not dependent on personal conditions or interpretations. This collective confirmation strengthens the reliability of scientific knowledge, promoting confidence in theories that withstand repeated empirical testing under similar conditions.

Q27. Answer: B

Explanation: Empirical evidence originates from systematic observation and measurement, forming the foundation of scientific inquiry. It provides verifiable data that can be tested and replicated. This reliance on observation distinguishes empirical science from speculative or theoretical claims lacking measurable support, ensuring objectivity and consistency in knowledge building.

Q28. Answer: C

Explanation: Reliability reflects the consistency of results when experiments are repeated under similar conditions. Reliable data exhibit stability, indicating minimal random error and reinforcing trust in observed patterns. This property ensures that findings are dependable and replicable, a hallmark of sound experimental design and credible scientific research.

Q29. Answer: B

Explanation: Validity determines whether an experiment accurately measures its intended concept or hypothesis. It ensures that conclusions legitimately reflect causal relationships, not confounding factors.

By designing precise variables and controls, researchers enhance both internal and external validity, yielding results that truly represent the phenomena being studied.

Q30. Answer: A

Explanation: Transparency enables other scientists to review, replicate, and validate data, ensuring accountability and reliability. Open access promotes cumulative knowledge and exposes potential errors. This practice safeguards against misconduct and fosters collaborative progress, upholding the scientific principle of openness and collective verification.

Q31. Answer: A

Explanation: The null hypothesis assumes no relationship between variables, providing a baseline for statistical comparison. The experimental hypothesis posits a predicted effect or difference. Testing the null hypothesis allows researchers to determine whether observed results are statistically significant, strengthening scientific conclusions through evidence-based inference.

Q32. Answer: C

Explanation: Qualitative observations describe characteristics or qualities without numerical measurement, such as color, texture, or odor. They provide valuable contextual information complementing quantitative data. Together, both forms of observation create a comprehensive understanding of experimental phenomena, balancing precision with descriptive insight.

Q33. Answer: B

Explanation: Statistical analysis enables objective interpretation of data by assessing variability, patterns, and significance. It determines whether observed effects are due to chance or genuine relationships. By quantifying uncertainty, statistics help validate hypotheses, ensuring that conclusions are robust and scientifically defensible across studies.

Q34. Answer: B

Explanation: Double-blind designs prevent both participants and researchers from knowing treatment assignments, reducing expectancy effects and bias. This approach maintains objectivity and ensures that outcomes reflect true experimental effects. By limiting conscious or unconscious influence, double-blind studies uphold the integrity of empirical testing.

Q35. Answer: D

Explanation: Precision reflects the consistency of repeated measurements, reducing uncertainty in data. High precision strengthens confidence in observed patterns and enhances reproducibility. While precision does not guarantee accuracy, its combination with calibrated instruments and proper technique ensures high-quality data fundamental to credible scientific investigation.

Q36. Answer: B

Explanation: Predictive power measures how effectively a model anticipates observable outcomes, validating its theoretical soundness. Models with strong predictive accuracy reflect correct underlying mechanisms and assumptions. Their success in forecasting experimental results demonstrates explanatory strength, guiding refinement and integration into broader scientific theory.

Q37. Answer: C

Explanation: Control variables are held constant to prevent interference with the relationship between independent and dependent variables. Their stability ensures that experimental changes result from deliberate manipulation rather than external variation, enhancing internal validity and clarity of causal inference within scientific investigations.

Q38. Answer: B

Explanation: Data interpretation translates numerical or qualitative observations into coherent explanations aligned with hypotheses. Statistical and logical analysis reveal trends and relationships, converting raw results into insights that advance understanding. This interpretive step bridges observation and theory, forming the basis for valid scientific conclusions.

Q39. Answer: B

Explanation: Negative results challenge assumptions and reveal limitations, guiding refinement of hypotheses and methods. Publishing such outcomes maintains transparency and prevents confirmation bias. They contribute equally to scientific progress by showing what does not work, preserving objectivity and enhancing cumulative understanding.

Q40. Answer: A

Explanation: External validity improves when results generalize across populations, conditions, and settings. Using large, representative samples increases applicability of findings beyond the experiment. By reflecting natural variability, such designs ensure that scientific conclusions are broadly relevant, enhancing their impact and reliability.

Q41. Answer: B

Explanation: Selective reporting introduces bias by emphasizing data that supports expectations while omitting contradictory findings. This distorts interpretations and undermines credibility. Ethical science requires transparency and balance, presenting all results objectively to ensure that conclusions genuinely reflect empirical evidence rather than preference.

Q42. Answer: B

Explanation: Reproducibility demonstrates that findings hold under independent conditions, confirming objectivity. It guards against error and fraud, enabling cumulative advancement of knowledge. This principle ensures science remains a shared, verifiable process where reliability, not authority, determines the acceptance of ideas and results.

Q43. Answer: C

Explanation: A testable hypothesis specifies measurable variables and observable outcomes, allowing empirical validation. The statement about meditation and heart rate can be tested through controlled observation. By defining variables and expected results, such hypotheses align with the scientific requirement for falsifiability and measurable evidence.

Q44. Answer: B

Explanation: Scientific skepticism involves questioning claims until supported by robust evidence. It safeguards objectivity, ensuring that theories rest on verifiable data rather than assumption or bias. This critical stance promotes accuracy and integrity, allowing science to evolve through evidence-based refinement and correction of past misconceptions.

Q45. Answer: B

Explanation: A placebo controls for psychological or expectation-based influences, isolating the true physiological effect of a treatment. By providing a comparison group, it enhances reliability and validity in medical or behavioral studies. Placebos ensure that measured effects result from the intervention itself, not from participant belief.

Q46. Answer: A

Explanation: Accuracy measures how close results are to the true or accepted value, while precision measures consistency across trials. High accuracy and precision together indicate reliable data. Distinguishing these concepts allows scientists to diagnose errors and improve measurement techniques in experimental research.

Q47. Answer: B

Explanation: Control conditions establish a reference baseline, allowing direct comparison with experimental treatments. By keeping all other variables constant, researchers can attribute observed differences to the independent variable. This structure enhances internal validity and clarifies cause-and-effect relationships within scientific investigations.

Q48. Answer: A

Explanation: Pseudoscience relies on untestable claims, anecdotal evidence, and resistance to revision, lacking empirical grounding. Unlike science, it avoids falsification and objective validation. Identifying pseudoscientific practices protects scientific integrity, ensuring that accepted theories derive from verifiable, systematic investigation, not belief or speculation.

Q49. Answer: B

Explanation: Science's defining trait is its reliance on testable, empirical evidence derived from observation and experimentation. This characteristic separates it from authority-based or subjective knowledge systems. By prioritizing verification and replicability, science ensures that understanding evolves through measurable and objective inquiry.

Q50. Answer: A

Explanation: Clear communication ensures that findings can be scrutinized, reproduced, and built upon by others. Transparent reporting of methods, data, and conclusions facilitates collaboration and innovation. This openness underpins science's cumulative nature, transforming individual discoveries into collective advancement of human knowledge.

Q51. Answer: D

Explanation: After experimentation, scientists analyze data and compare outcomes to the original hypothesis to determine support or refutation. This evaluative stage connects empirical results with theoretical expectations. It ensures conclusions are evidence-driven, guiding refinement of hypotheses or models and contributing to the iterative, self-correcting nature of science.

Q52. Answer: A

Explanation: Standardized units, such as SI, allow scientists worldwide to communicate results accurately and compare findings. They eliminate ambiguity and ensure consistent data interpretation. This uniformity underpins collaboration, reproducibility, and integration of research across disciplines, facilitating the global advancement of scientific knowledge.

Q53. Answer: B

Explanation: Variables are factors that can vary, influencing experimental outcomes. They include independent, dependent, and controlled types. Understanding how variables interact is essential for establishing cause-and-effect relationships and ensuring clarity, consistency, and validity in scientific experimentation and analysis.

Q54. Answer: A

Explanation: Models are dynamic tools updated as new data and insights emerge. Revisions enhance accuracy and explanatory power, aligning theoretical constructs with observed phenomena. This adaptability demonstrates science's progressive refinement and responsiveness to empirical validation through ongoing experimentation and discovery.

Q55. Answer: B

Explanation: Creativity fuels innovation by helping scientists formulate hypotheses, design experiments, and interpret data in novel ways. While grounded in logic and evidence, scientific creativity expands conceptual frameworks and fosters breakthroughs. It integrates imagination with critical thinking, driving advancement in methodology and theoretical development.

Q56. Answer: C

Explanation: Basic research seeks to expand understanding of fundamental principles without immediate practical goals. It forms the foundation for applied science by uncovering mechanisms that later inform technology or solutions. Its emphasis on curiosity-driven discovery strengthens the theoretical base for future innovation and application.

Q57. Answer: A

Explanation: The scientific method addresses empirical phenomena but cannot resolve moral, ethical, or aesthetic judgments, which lie outside empirical testability. Its scope is confined to observable, measurable domains. Recognizing these boundaries maintains the integrity and focus of scientific inquiry on natural, verifiable processes.

Q58. Answer: C

Explanation: A literature review compiles and critiques existing research to identify knowledge gaps, trends, and frameworks. It informs hypothesis formulation and methodological design, ensuring new investigations build upon established findings. This process reinforces continuity, preventing redundancy and enhancing the coherence of scientific inquiry.

Q59. Answer: A

Explanation: The reproducibility crisis arises when independent researchers cannot replicate published findings, casting doubt on reliability. It highlights methodological flaws, biases, or insufficient transparency. Addressing this issue through data sharing, open methods, and rigorous replication strengthens trust and credibility within the scientific community.

Q60. Answer: B

Explanation: Applied research focuses on solving practical problems, such as vaccine development, using existing scientific knowledge. It translates theoretical insights from basic research into real-world applications. This pragmatic approach bridges the gap between discovery and societal benefit, demonstrating science's direct impact on human welfare.

Q61. Answer: A

Explanation: Blind procedures prevent participants or experimenters from knowing critical details, minimizing expectancy effects. This design enhances objectivity and data validity by removing psychological influence. Blinding ensures that outcomes reflect genuine effects rather than biases, reinforcing the integrity of scientific experimentation.

Q62. Answer: A

Explanation: Uncertainty quantifies possible error in measurement, defining precision and reliability. Acknowledging uncertainty allows scientists to assess confidence in results and compare datasets meaningfully. This transparency ensures that scientific conclusions account for variability, strengthening the robustness of empirical interpretation.

Q63. Answer: B

Explanation: A paradigm shift, as described by Thomas Kuhn, occurs when new evidence or theories fundamentally transform existing scientific frameworks. It redefines accepted assumptions and methodologies, exemplifying science's evolving nature. Such revolutions, like the shift from Newtonian to Einsteinian physics, mark profound conceptual advancement.

Q64. Answer: A

Explanation: Peer review and replication confirm the credibility of data through expert scrutiny and independent validation. They detect errors and assess methodological rigor, ensuring accuracy and reliability. This collaborative process underpins scientific integrity, reinforcing confidence in evidence-based conclusions across disciplines.

Q65. Answer: B

Explanation: Correlation reveals statistical relationships between variables, while causation demonstrates direct influence. Confusing the two leads to misinterpretation of data. Establishing causality requires controlled experimentation and evidence of mechanism, ensuring that conclusions accurately reflect underlying relationships in nature.

Q66. Answer: D

Explanation: Integrity requires honesty in data collection, analysis, and reporting. Fabrication or falsification violates ethical standards, undermining trust and credibility. Upholding integrity ensures that findings accurately represent reality and reinforces the collective reliability of the scientific record within the research community.

Q67. Answer: A

Explanation: Randomization ensures that participants are assigned without bias, equalizing group characteristics. It eliminates systematic error and strengthens internal validity. This practice makes results more generalizable and minimizes confounding factors, thereby enhancing the fairness and accuracy of scientific conclusions.

Q68. Answer: B

Explanation: The null hypothesis offers a neutral baseline suggesting no effect or difference, allowing statistical analysis to assess significance. Rejection or retention of the null provides objective evidence regarding relationships among variables, ensuring decisions are grounded in quantifiable data rather than assumption or bias.

Q69. Answer: A

Explanation: Replication tests whether results hold under similar conditions, verifying reliability or revealing limitations. It maintains the self-correcting cycle of science, distinguishing genuine discoveries from anomalies. Replication consolidates trust in theories through continuous testing and validation across independent investigations.

Q70. Answer: A

Explanation: Calibration aligns instruments with known standards, ensuring measurement accuracy and consistency. Regular calibration corrects systematic errors and maintains reliability across experiments. This process upholds data quality, supporting valid conclusions and comparability in scientific research.

Q71. Answer: B

Explanation: Abductive reasoning infers the most likely explanation for observations, integrating inductive pattern recognition with deductive logic. It underlies hypothesis generation, bridging data and theory formation. This reasoning type enhances creative problem-solving in science, fostering development of plausible, testable scientific models.

Q72. Answer: C

Explanation: Detailed methodology ensures transparency, allowing others to replicate procedures and verify findings. Clear documentation supports accountability and reproducibility, which are essential for validation. Publishing methods empowers collaborative review, strengthening confidence in scientific accuracy and reliability.

Q73. Answer: B

Explanation: Open-mindedness encourages scientists to revise theories when confronted with new evidence, ensuring adaptability. It maintains science as a dynamic, evolving process guided by data rather than dogma. This attitude fosters innovation, self-correction, and progressive refinement of scientific understanding.

Q74. Answer: A

Explanation: A pilot study evaluates design, logistics, and procedures before large-scale implementation. It identifies potential problems, improving validity and efficiency of subsequent research. Pilot testing enhances methodological soundness, minimizing errors and optimizing resources in comprehensive experimental work.

Q75. Answer: B

Explanation: Science is a dynamic enterprise grounded in empirical evidence and continual revision. It evolves through observation, testing, and theoretical refinement. Openness to change ensures that knowledge remains current, accurate, and reflective of reality, exemplifying the self-correcting and progressive nature of scientific understanding.

Q76. Answer: D

Explanation: Velocity is a vector quantity because it has both magnitude and direction, unlike speed which only measures magnitude. Understanding this distinction is crucial for motion analysis, as direction affects displacement outcomes. Distance and work are scalars, lacking directional properties. Power represents the rate of energy transfer, also scalar. Recognizing vector versus scalar quantities helps in applying Newton's laws correctly across various motion scenarios.

Q77. Answer: C

Explanation: Even at constant speed, circular motion involves changing direction, producing centripetal acceleration toward the center. This inward acceleration maintains the curved path by continuously altering velocity's direction. Speed remains constant but velocity changes because direction changes. Outward acceleration would imply an unrealistic centrifugal force. Kinetic energy remains constant because speed is constant. Understanding this explains orbital motion and rotating system dynamics.

Q78. Answer: C

Explanation: Using Newton's Second Law, acceleration equals force divided by mass ($a = F/m$). Here, 10 N / 2 kg = 5 m/s^2. The acceleration's direction matches the net force direction, showing proportionality between applied force and resulting motion. Understanding this principle allows prediction of movement under varying forces. This simple relationship forms the foundation of dynamics and mechanical problem solving.

Q79. Answer: A

Explanation: Newton's Third Law states that every action has an equal and opposite reaction. In rocket propulsion, gases expelled downward exert an equal upward force on the rocket, causing lift. This reciprocal interaction illustrates the symmetry of forces between interacting bodies. The other examples describe motion but not paired force interactions. Recognizing this helps explain thrust, recoil, and propulsion systems.

Q80. Answer: C

Explanation: Momentum equals mass multiplied by velocity ($p = mv$), representing an object's motion quantity. It is a vector quantity, having direction and magnitude. This definition underlies conservation of momentum principles in collisions and interactions. Force over time changes momentum, linking it to impulse. Understanding momentum clarifies both macroscopic impacts and subatomic particle collisions in physical systems.

Q81. Answer: D

Explanation: Kinetic energy equals $\frac{1}{2}mv^2$; when velocity doubles, the square of velocity quadruples the energy. Thus, a twofold speed increase yields four times the kinetic energy, illustrating the exponential relation between velocity and energy. This principle explains why small velocity increases lead to large energy changes in vehicles, projectiles, and atomic particles, emphasizing energy control in mechanics.

Q82. Answer: B

Explanation: Work equals the product of applied force and displacement in the same direction ($W = F \cdot d \cdot \cos\theta$). It measures energy transfer through mechanical means. Only force components parallel to displacement perform work. This concept links force, motion, and energy conservation. Recognizing the directionality of force relative to motion is essential for analyzing machines and physical processes.

Q83. Answer: C

Explanation: Elastic collisions conserve both momentum and kinetic energy because no energy converts to heat or deformation. This idealization models atomic and molecular interactions. Real-world collisions are partially elastic, but the principle still approximates energy transfer accurately. Understanding these conservation laws is central to analyzing impact mechanics and subatomic scattering experiments.

Q84. Answer: C

Explanation: The slope of a velocity-time graph equals acceleration, showing the rate of velocity change. A positive slope means speeding up; a negative slope indicates slowing down. This visual method provides intuitive understanding of motion and is fundamental in kinematic graph interpretation. It helps translate between mathematical descriptions and real-world motion analysis effectively.

Q85. Answer: A

Explanation: The first law asserts conservation of energy, meaning total energy remains constant though it changes form between heat, work, and internal energy. This foundational principle connects mechanics and thermodynamics. Understanding it clarifies how energy transfer underpins engines, biological processes, and physical transformations across natural systems.

Q86. Answer: B

Explanation: Power measures the rate of energy transfer or work per unit time, with SI unit watt ($1\ W = 1\ J/s$). It quantifies how quickly work is done or energy is consumed. Recognizing power units helps evaluate performance of engines, machines, and electrical systems. It links energy, time, and efficiency in real-world applications.

Q87. Answer: C

Explanation: Satellites remain in orbit because their tangential velocity balances gravitational attraction, creating continuous free-fall around Earth. They move forward fast enough that the surface curves away beneath them. This balance between inertia and gravity sustains orbital motion. Understanding orbital dynamics connects Newtonian mechanics with modern spaceflight principles and celestial motion stability.

Q88. Answer: D

Explanation: Electromagnetic waves consist of perpendicular oscillating electric and magnetic fields that propagate through space without a medium. Their speed in vacuum is constant (3×10^8 m/s). They transmit energy, not matter. This concept unifies light, radio, and X-rays, illustrating Maxwell's synthesis of electromagnetism and wave theory fundamental to physics and technology.

Q89. Answer: C

Explanation: Wave energy is proportional to the square of its amplitude, showing that doubling amplitude quadruples energy. Amplitude measures displacement intensity; higher amplitude means greater energy transport. This relation holds in mechanical and electromagnetic waves, illustrating energy concentration in sound, light, and seismic vibrations used in engineering and physics analysis.

Q90. Answer: B

Explanation: Ohm's Law states $I = V/R$; current increases with applied voltage but decreases with resistance. It models electron flow through conductive materials. Understanding this relationship is vital for analyzing electrical circuits and designing systems for safe, efficient operation in electronics and power distribution.

Q91. Answer: C

Explanation: A simple pendulum converts energy between kinetic and potential forms; at lowest point kinetic energy is maximal while potential energy is minimal, and vice versa at highest points. In absence of nonconservative forces total mechanical energy remains constant, neglecting air resistance and friction. Amplitude affects maximum potential energy but not period for small angles. This energy interchange exemplifies conservation principles in oscillatory motion and informs real-world pendulum engineering design practice.

Q92. Answer: C

Explanation: Projectile motion neglecting air resistance separates horizontal and vertical components; horizontal velocity remains constant while vertical velocity changes under constant acceleration due to gravity. The range depends on launch speed and angle, maximized at forty-five degrees for level ground when starting and landing heights are equal. Understanding component independence simplifies solving trajectories and enables prediction of landing positions, peak height, and flight time using kinematic equations and practical trajectory design.

Q93. Answer: C

Explanation: Centripetal force is the net inward force required to keep an object moving in a circular path, with magnitude mv^2/r directed toward the center. It arises from tension, gravity, friction, or normal forces depending on the context. Without sufficient centripetal force the object moves tangentially off the curve. Recognizing centripetal versus centrifugal descriptions prevents conceptual errors when analyzing rotating systems and applying Newton's laws to circular motion in engineering analysis.

Q94. Answer: B

Explanation: In an explosion or separation event internal forces change individual object momenta but total momentum of the isolated system remains conserved because no external impulse acts. Fragments move with velocities such that vector sum of their momenta equals pre-event momentum. Conservation allows solving for unknown velocities post-separation. This principle underlies recoil of firearms and fragmentation dynamics, and it is essential for analyzing collisions and separations in closed systems in practice.

Q95. Answer: A

Explanation: Hydrostatic pressure at a given depth in a fluid increases with depth, described by $p = \rho gh$ for incompressible fluids, where ρ is density, g is gravitational acceleration, and h is depth. Pressure acts isotropically, transmitting forces equally in all directions at a point. Surface area affects force magnitude but not pressure. These concepts explain buoyancy, pressure gauges, and why deep-sea environments exert extreme forces on structures and organisms today.

Q96. Answer: B

Explanation: The Doppler effect causes apparent frequency shifts when source or observer moves relative to the medium: approaching motion increases observed frequency and receding motion decreases it. For sound this depends on velocities relative to the medium, requiring distinct formulas for moving source or observer. It is used in radar, medical imaging, and astronomy (redshift/blueshift analogues). Understanding sign conventions and frame-of-reference clarifies observed pitch changes in everyday and scientific contexts applications.

Q97. Answer: C

Explanation: Photon energy equals Planck's constant times frequency ($E=hf$), linking electromagnetic wave frequency to quantized energy packets. Higher-frequency photons like ultraviolet and X-rays carry more energy per photon than visible or radio waves, explaining ionization potential and material interactions. This quantum relationship anchors photoelectric effect explanations, spectral analysis, and energy transfer at microscopic scales, bridging classical wave descriptions and quantum particle behavior in modern physics applications used in spectroscopy and imaging.

Q98. Answer: C

Explanation: In an RC circuit the time constant $\tau = RC$ defines the characteristic charge and discharge time of the capacitor through the resistor, representing the interval over which voltages change by about 63.2% of their difference. After multiple time constants the system approaches steady state exponentially. Understanding τ aids circuit timing design, filters, and transient response analysis in electronics, controlling signal shaping, sensor response, and simple timing-based applications in practice.

Q99. Answer: B

Explanation: Diffraction limits the resolving power of optical systems because waves diffract when passing apertures, producing interference patterns; the smallest resolvable angle is roughly λ/D for a circular aperture, where λ is wavelength and D aperture diameter. Increasing aperture size or using shorter wavelengths improves resolution. This principle governs telescope and microscope design, explains Airy disk formation, and sets fundamental limits on imaging systems and instruments across scientific and engineering disciplines.

Q100. Answer: B

Explanation: Bernoulli's principle relates fluid pressure and flow speed along streamlines: faster flow corresponds to lower pressure, producing pressure differentials that contribute to lift on airfoils. However, wing lift also depends on circulation, angle of attack, and Newtonian action–reaction deflecting air downward. Realistic lift explanations combine Bernoulli and Newtonian perspectives, emphasizing streamlined shapes and pressure distributions in aerodynamic design for aircraft and wind turbine blades.

Questions (101-200)

Q101. The work-energy theorem states that the net work done on an object equals:

A. The change in its kinetic energy
B. The product of force and acceleration
C. The change in its potential energy
D. Its momentum multiplied by distance
E. The total mechanical energy

Q102. When current flows through a resistor, electrical energy is converted into:

A. Magnetic energy
B. Kinetic energy
C. Heat energy
D. Chemical energy
E. Potential energy

Q103. A transformer operates on which physical principle?

A. Conservation of charge
B. Electrostatic induction
C. Electromagnetic induction
D. Photoelectric effect
E. Thermal conduction

Q104. The mechanical advantage of a lever increases when:

A. The effort arm shortens
B. The load arm lengthens
C. The effort arm lengthens relative to load arm
D. Both arms are equal length
E. The fulcrum is centered

Q105. Which of the following particles experiences the largest deflection in a magnetic field?

A. A slow, massive particle
B. A fast, heavy ion
C. A light, charged particle moving perpendicularly
D. A neutral atom
E. A stationary electron

Q106. Which of the following best describes potential difference?

A. The total charge in a circuit
B. The work done per unit charge moving between two points
C. The product of resistance and current squared
D. The energy stored in a capacitor
E. The sum of all currents in a loop

Q107. The rate at which charge flows through a surface is defined as:

A. Resistance
B. Current
C. Power
D. Voltage
E. Capacitance

Q108. The period of a simple pendulum depends primarily on:

A. The mass of the bob
B. The amplitude of swing
C. The length of the pendulum and gravity

D. The material of the string
E. The air resistance

Q109. The resistance of a wire increases when:
A. Its temperature decreases
B. Its length decreases
C. Its cross-sectional area increases
D. Its temperature increases
E. It is cooled to zero Kelvin

Q110. According to Coulomb's Law, the force between two charges:
A. Is inversely proportional to the product of charges
B. Is inversely proportional to the square of the distance
C. Increases linearly with distance
D. Is independent of the medium
E. Is always attractive

Q111. Magnetic flux through a surface depends on:
A. The magnetic field strength and orientation to the surface
B. The temperature of the magnet
C. The electric field strength
D. The velocity of the conductor
E. The resistance of the wire

Q112. In harmonic motion, resonance occurs when:
A. The system is overdamped
B. The driving frequency equals natural frequency
C. The driving force is zero
D. The amplitude decreases over time
E. Frictional forces dominate motion

Q113. The capacitance of a parallel-plate capacitor increases when:
A. Plate area decreases
B. Plate separation increases
C. A dielectric is removed
D. Plate area increases or separation decreases
E. Voltage decreases

Q114. In simple harmonic motion, acceleration is:
A. Constant and directed upward
B. Proportional to displacement and opposite in direction
C. Inversely proportional to velocity
D. Constant and directed downward
E. Always zero at mean position

Q115. Which law explains buoyant force acting on a submerged object?
A. Bernoulli's law
B. Archimedes' principle
C. Pascal's law
D. Newton's second law
E. Hooke's law

Q116. Which physical quantity is conserved in all closed systems regardless of type of collision?
A. Kinetic energy
B. Potential energy
C. Momentum

D. Elastic energy
E. Rotational energy

Q117. The total mechanical energy of a system remains constant when:

A. Nonconservative forces do work
B. Only conservative forces act
C. Kinetic energy increases continuously
D. Friction acts on the body
E. External forces are applied

Q118. When a gas expands adiabatically:

A. Heat is absorbed from surroundings
B. Temperature remains constant
C. No heat is exchanged with surroundings
D. Pressure increases
E. Volume decreases

Q119. The electric field inside a conductor in electrostatic equilibrium is:

A. Zero
B. Constant nonzero
C. Equal to applied field
D. Maximum at surface
E. Random

Q120. Which of the following properties of waves is independent of the medium?

A. Speed
B. Frequency
C. Wavelength
D. Amplitude
E. Intensity

Q121. The magnetic field at the center of a circular loop of radius r carrying current I is:

A. $\mu_0 I/(2\pi r)$
B. $\mu_0 I/(4\pi r^2)$
C. $\mu_0 I/(2r)$
D. $\mu_0 I^2/r$
E. $\mu_0 I/\pi r^2$

Q122. In a standing wave, nodes are points where:

A. Displacement is maximum
B. Displacement is zero
C. Velocity is maximum
D. Pressure is constant
E. Energy transfer is maximum

Q123. The pressure in a fluid at rest:

A. Acts only downward
B. Acts equally in all directions
C. Depends on container shape
D. Is zero at all depths
E. Acts only upward

Q124. When two waves of slightly different frequencies interfere, the phenomenon produced is:

A. Resonance
B. Diffraction
C. Beats

D. Reflection
E. Refraction

Q125. The Lorentz force on a charged particle moving in electric (E) and magnetic (B) fields is:

A. qE
B. qB
C. q(E + v × B)
D. q(E – vB)
E. qv/E

Q126. A 5-kg object moving at 4 m/s collides elastically with a stationary 3-kg object. The total momentum after collision is:

A. 0 kg·m/s
B. 12 kg·m/s
C. 20 kg·m/s
D. 15 kg·m/s
E. 30 kg·m/s

Q127. A heat engine with 400 J input performs 100 J of work. What is its efficiency?

A. 10%
B. 25%
C. 50%
D. 75%
E. 100%

Q128. A material obeying Hooke's law exhibits:

A. Constant stress at all strains
B. Proportionality between stress and strain
C. Nonlinear stress-strain relation
D. Permanent deformation under load
E. Increasing stiffness with time

Q129. Which of the following statements about photons is true?

A. They have mass but no charge
B. They have charge but no mass
C. They have neither mass nor charge but carry energy and momentum
D. They have mass and charge
E. They have variable rest mass

Q130. The escape velocity from a planet depends on:

A. Its density only
B. Its mass and radius
C. The object's mass
D. The planet's orbital velocity
E. Atmospheric conditions

Q131. The second law of thermodynamics states:

A. Energy can be created in isolated systems
B. Entropy of an isolated system never decreases
C. Work is independent of heat flow
D. All heat can be converted to work
E. Systems naturally become more ordered

Q132. Which of the following particles has the greatest charge-to-mass ratio?

A. Proton
B. Electron

C. Neutron
D. Alpha particle
E. Deuteron

Q133. The principle of superposition applies when:

A. Forces act in only one direction
B. The system behaves linearly
C. The system exhibits nonlinear responses
D. Only one force acts at a time
E. Objects are in equilibrium

Q134. The wavelength of a sound wave increases when:

A. Frequency increases at constant speed
B. Speed increases at constant frequency
C. Speed decreases at constant frequency
D. Both speed and frequency decrease
E. Amplitude increases

Q135. The principle of conservation of angular momentum explains:

A. Why a spinning skater speeds up when arms are pulled in
B. Why objects fall at constant acceleration
C. Why light refracts at an interface
D. Why heat transfers spontaneously
E. Why resistance increases with temperature

Q136. A semiconductor's conductivity increases when:

A. Temperature decreases
B. Temperature increases
C. It is cooled below 0°C
D. Voltage decreases
E. Magnetic field increases

Q137. The relationship between pressure and volume at constant temperature is given by:

A. Boyle's law
B. Charles's law
C. Gay-Lussac's law
D. Avogadro's law
E. Dalton's law

Q138. The SI unit of electric charge is:

A. Coulomb
B. Ampere
C. Volt
D. Ohm
E. Tesla

Q139. Which color of visible light has the highest energy photons?

A. Red
B. Orange
C. Yellow
D. Green
E. Violet

Q140. The magnitude of the electric field between two parallel plates separated by distance d and potential difference V is:

A. Vd

B. V/d
C. d/V
D. V^2/d
E. 1/(Vd)

Q141. When an object moves in a viscous fluid at constant velocity, the net force on it is:
A. Downward
B. Upward
C. Zero
D. Equal to its weight
E. Equal to its drag force

Q142. The restoring force in a spring system is described by:
A. F = ma
B. F = -kx
C. $F = kx^2$
D. F = mg
E. F = qE

Q143. When white light passes through a prism, dispersion occurs because:
A. Light slows equally for all colors
B. Refractive index depends on wavelength
C. Reflection increases with angle
D. Energy decreases for shorter wavelengths
E. Colors recombine at exit

Q144. The unit of inductance is:
A. Farad
B. Ohm
C. Henry
D. Weber
E. Tesla

Q145. The refractive index of a material is defined as:
A. Ratio of speed of light in vacuum to that in material
B. Ratio of wavelength in material to that in vacuum
C. Product of light speed and density
D. Inverse of optical density
E. Constant independent of wavelength

Q146. The energy stored in a stretched spring is:
A. $\frac{1}{2}kx^2$
B. kx
C. kx^2
D. $2kx^2$
E. x/k

Q147. In nuclear fission:
A. Small nuclei combine to form larger ones
B. Large nuclei split into smaller nuclei
C. Energy is absorbed
D. No mass changes occur
E. Only light elements are involved

Q148. In an RLC circuit at resonance:
A. Current is minimum

B. Voltage leads current by 90°
C. Inductive and capacitive reactances cancel
D. Power factor is zero
E. Resistance is infinite

Q149. Which gas law combines pressure, volume, and temperature relationships?
A. Boyle's law
B. Charles's law
C. Combined gas law
D. Avogadro's law
E. Ideal gas law

Q150. A concave mirror produces a real image when the object is:
A. At the focus
B. Between focus and mirror
C. Beyond the center of curvature
D. At infinity only
E. At the center of curvature

Q151. The acceleration due to gravity on Earth varies slightly because:
A. The atmosphere changes density
B. Earth's mass changes daily
C. Earth's shape and rotation cause differences
D. Gravity is stronger at equator
E. Gravity is uniform everywhere

Q152. In an ideal gas, average kinetic energy of molecules depends on:
A. Pressure
B. Volume
C. Temperature
D. Number of molecules
E. Density

Q153. The escape of molecules from a liquid surface into the air is called:
A. Condensation
B. Sublimation
C. Evaporation
D. Deposition
E. Melting

Q154. A sound wave cannot travel through:
A. Air
B. Water
C. Steel
D. Vacuum
E. Wood

Q155. If a current-carrying conductor is placed parallel to a magnetic field, the force on it is:
A. Maximum
B. Zero
C. Minimum but not zero
D. Perpendicular to the current
E. Directed toward the field source

Q156. The half-life of a radioactive substance is the time required for:
A. All atoms to decay completely

B. Half of the sample to decay
C. Activity to double
D. Activity to remain constant
E. Temperature to halve

Q157. The property of a material that opposes magnetic field formation within it is called:

A. Permeability
B. Permittivity
C. Resistivity
D. Reluctance
E. Diamagnetism

Q158. The phase difference between current and voltage in a purely inductive circuit is:

A. 0°
B. 30°
C. 45°
D. 60°
E. 90°

Q159. The buoyant force on an object immersed in a fluid depends on:

A. The object's shape
B. The object's density
C. The fluid's density and displaced volume
D. The object's surface area only
E. Atmospheric pressure

Q160. The total internal reflection of light occurs when:

A. Light passes from rarer to denser medium
B. Incidence angle is less than critical angle
C. Light passes from denser to rarer medium at angle greater than critical angle
D. Light travels in a straight line
E. Surface is rough

Q161. Which instrument measures electric current directly?

A. Voltmeter
B. Ammeter
C. Ohmmeter
D. Wattmeter
E. Galvanometer

Q162. The magnetic field lines around a straight current-carrying conductor are:

A. Straight and parallel
B. Circular and concentric
C. Elliptical
D. Divergent lines
E. Radial

Q163. The potential energy of a satellite in orbit is:

A. Positive and increases with altitude
B. Negative and increases (becomes less negative) with altitude
C. Zero everywhere
D. Positive and constant
E. Negative and constant

Q164. The color of light emitted by an atom depends on:

A. The atom's velocity

B. Energy differences between electron levels
C. Temperature of surroundings
D. Light intensity
E. Electric field strength

Q165. The unit of electric capacitance is:

A. Volt
B. Henry
C. Farad
D. Coulomb
E. Ohm

Q166. The magnitude of centripetal acceleration is proportional to:

A. Radius of the circle
B. Square of speed and inversely to radius
C. Radius squared
D. Speed only
E. Constant mass

Q167. In an ideal transformer, power output equals:

A. Power input minus heat losses
B. Power input
C. Zero
D. Product of primary current and secondary voltage only
E. Power input plus core losses

Q168. The unit of frequency is:

A. Joule
B. Hertz
C. Watt
D. Volt
E. Ohm

Q169. A concave lens always produces:

A. Real and inverted images
B. Virtual and upright images
C. Real and magnified images
D. Virtual and inverted images
E. Real and upright images

Q170. When electrons move through a conductor, they lose energy primarily as:

A. Heat
B. Sound
C. Light
D. Magnetic field energy
E. Electric potential

Q171. Which of the following statements about acceleration due to gravity is true?

A. It depends on the falling object's mass
B. It decreases slightly with height above Earth
C. It increases with altitude
D. It is constant throughout the universe
E. It depends only on object's volume

Q172. The mechanical energy of a pendulum is maximum when:

A. At highest point

B. At mean position
C. At either end of swing
D. When acceleration is zero
E. When potential energy equals kinetic energy

Q173. The slope of a distance-time graph represents:
A. Velocity
B. Acceleration
C. Displacement
D. Jerk
E. Momentum

Q174. The energy of an electromagnetic wave is equally divided between:
A. Magnetic field only
B. Electric field only
C. Electric and magnetic fields
D. Photons and electrons
E. Charge and mass

Q175. The heat required to raise the temperature of 1 kg of a substance by 1°C is called:
A. Latent heat
B. Specific heat capacity
C. Thermal conductivity
D. Heat of fusion
E. Molar heat

Q176. The intensity of sound decreases with distance because:
A. Sound speed decreases
B. Energy spreads over a larger area
C. Air absorbs all sound energy instantly
D. Wavelength increases
E. Frequency decreases

Q177. The electric field lines between two opposite charges:
A. Are circular
B. Are parallel
C. Originate from positive and terminate on negative charge
D. Form closed loops
E. Point from negative to positive charge

Q178. When light passes from air into glass, which changes occur?
A. Frequency decreases, speed increases
B. Frequency constant, speed decreases
C. Both speed and frequency decrease
D. Speed constant, wavelength increases
E. Frequency increases, wavelength constant

Q179. The primary purpose of a fuse in an electrical circuit is to:
A. Reduce current flow
B. Store charge
C. Protect against excessive current
D. Increase voltage
E. Convert current to heat efficiently

Q180. The torque on a dipole in a uniform electric field is maximum when the dipole:
A. Is parallel to the field

B. Is perpendicular to the field
C. Is anti-parallel to the field
D. Makes a 45° angle with the field
E. Is at equilibrium

Q181. A transformer's primary and secondary coils are linked by:
A. Electric conduction
B. Mechanical contact
C. Magnetic flux
D. Static charge
E. Direct connection

Q182. The kinetic energy of a rotating body is:
A. $\frac{1}{2} I\omega^2$
B. $I\omega$
C. $\frac{1}{2} m\omega^2$
D. $\frac{1}{2} mv^2$
E. $I^2\omega$

Q183. If a gas is compressed at constant temperature, its pressure:
A. Decreases
B. Increases
C. Remains constant
D. Drops to zero
E. Depends on gas type only

Q184. The charge of an electron is approximately:
A. 1.6×10^{-19} C
B. 9.1×10^{-31} C
C. 3×10^{8} C
D. 1 C
E. 6.02×10^{23} C

Q185. The phenomenon responsible for blue color of the sky is:
A. Reflection
B. Scattering
C. Refraction
D. Diffraction
E. Dispersion

Q186. The motion of planets around the Sun is due to:
A. Centrifugal force
B. Electromagnetic attraction
C. Gravitational force
D. Magnetic force
E. Inertia alone

Q187. The thermal expansion of solids occurs because:
A. Molecules attract more strongly when heated
B. Intermolecular spaces increase with temperature
C. Molecules become smaller
D. Volume decreases with heat
E. Density increases

Q188. The magnetic effect of electric current was first discovered by:
A. Newton

B. Faraday
C. Ampère
D. Oersted
E. Coulomb

Q189. The time period of a wave is:

A. Number of oscillations per second
B. Inverse of frequency
C. Product of frequency and wavelength
D. Dependent on wave amplitude only
E. Equal to velocity over wavelength

Q190. The phenomenon of light bending around corners is known as:

A. Reflection
B. Refraction
C. Diffraction
D. Scattering
E. Interference

Q191. The kinetic theory of gases assumes that gas molecules:

A. Are large and stationary
B. Attract strongly at all distances
C. Move randomly with elastic collisions
D. Lose energy upon collision
E. Occupy significant volume

Q192. The electromagnetic spectrum is arranged according to:

A. Frequency or wavelength
B. Speed of light variation
C. Photon mass
D. Intensity
E. Temperature

Q193. Which of the following has the highest refractive index?

A. Air
B. Water
C. Glass
D. Diamond
E. Ice

Q194. The magnetic field strength around a solenoid increases when:

A. Current decreases
B. Fewer turns per unit length
C. Core is air instead of iron
D. Current and number of turns increase
E. Length increases

Q195. When a body floats, the weight of the body equals:

A. Its volume
B. Buoyant force
C. Fluid density
D. Pressure difference
E. Gravitational acceleration

Q196. The resistance of a conductor depends on:

A. Temperature, length, and cross-sectional area

B. Temperature only
C. Voltage only
D. Frequency only
E. Power only

Q197. The electric flux through a closed surface enclosing charge q is:

A. Zero
B. q/ε_0
C. ε_0/q
D. q^2/ε_0
E. Depends on surface shape

Q198. The potential energy of an object of mass m at height h is:

A. mgh
B. $\frac{1}{2}mv^2$
C. $\frac{1}{2}kx^2$
D. Fd
E. $\frac{1}{2}mgh$

Q199. In a photoelectric experiment, increasing light intensity increases:

A. Electron energy
B. Electron number emitted
C. Frequency threshold
D. Stopping potential
E. Work function

Q200. The slope of a stress-strain graph in the elastic region represents:

A. Elastic limit
B. Yield strength
C. Young's modulus
D. Breaking stress
E. Toughness

Answer keys and Explanations (101-200)

Q101. Answer: A

Explanation: The work-energy theorem links force and motion, stating that net work done on an object equals its change in kinetic energy. This relationship summarizes Newton's laws in energy terms, highlighting how forces transfer energy to accelerate or decelerate masses. Applying this theorem simplifies solving dynamic problems and connects work, energy, and force concepts, providing a powerful analytical tool across classical mechanics and engineering applications involving moving systems and mechanical power computations.

Q102. Answer: C

Explanation: Electric current through a resistor converts electrical energy into heat due to collisions between electrons and atomic lattice ions, described by Joule's law ($P=I^2R$). This process underlies heating elements and energy loss in circuits. Understanding this conversion is crucial for power management and safety in electrical systems. The principle links thermodynamics and electricity, demonstrating how energy transforms but remains conserved across physical processes encountered in engineering and daily technology.

Q103. Answer: C

Explanation: Transformers use electromagnetic induction to transfer energy between circuits through changing magnetic flux in a core. Alternating current in the primary coil generates varying magnetic fields inducing voltage in the secondary coil. The ratio of turns determines voltage change. This principle underpins power transmission efficiency, allowing step-up and step-down conversions while conserving power (neglecting losses) and illustrating Faraday's law applications in energy distribution networks worldwide in technology.

Q104. Answer: C

Explanation: Mechanical advantage equals the ratio of effort arm to load arm lengths. Increasing the effort arm relative to the load arm amplifies force output, allowing smaller input forces to lift heavier loads. This geometric relationship defines leverage principles used in tools, construction, and machinery. Recognizing torque relationships and balance points enables efficient design of mechanical systems that maximize force multiplication and reduce human or motor effort in practical applications.

Q105. Answer: C

Explanation: Magnetic force depends on charge, velocity, and magnetic field direction via $F=qvB\sin\theta$. A light, charged particle moving perpendicularly to the magnetic field undergoes maximum deflection due to low mass and high acceleration. Massive or stationary particles experience smaller or no deflection. Understanding this behavior informs mass spectrometry, cathode-ray tube operation, and particle physics experiments involving charged particle motion under magnetic influence.

Q106. Answer: B

Explanation: Potential difference, or voltage, equals the work done per unit charge moving between two points, measured in volts. It quantifies the electric potential energy available to drive charge flow. This concept links energy and electric fields, underlying circuit analysis and energy transfer efficiency.

Understanding potential difference is vital for designing circuits, measuring battery performance, and ensuring safe operation in electrical engineering and physics applications daily.

Q107. Answer: B

Explanation: Electric current is the rate of charge flow through a conductor's cross-section, measured in amperes (1 A = 1 C/s). It describes electron drift in response to potential differences. Recognizing current as charge flow helps analyze circuits and conservation laws (Kirchhoff's). Understanding current flow patterns is essential for circuit design, electronics troubleshooting, and energy transmission efficiency in practical electrical engineering systems and power supply analysis.

Q108. Answer: C

Explanation: The pendulum period $T=2\pi\sqrt{(L/g)}$ depends on its length and gravitational acceleration, but not on mass or amplitude (for small angles). Longer pendulums swing more slowly, illustrating harmonic motion properties. This principle underpins timing devices, seismographs, and oscillation-based instruments. Understanding period relationships strengthens comprehension of oscillatory systems and gravitational field influence in real-world measurement and design scenarios in physics and engineering disciplines.

Q109. Answer: D

Explanation: Resistance generally rises with temperature in metals because increased atomic vibrations hinder electron flow. The relationship follows $R = R_0(1+\alpha\Delta T)$, where α is temperature coefficient. Materials like semiconductors behave oppositely. Understanding this dependence informs material selection, thermal management, and design of resistive components. It also explains why overheating can damage electronics, emphasizing thermal regulation and safe current handling in circuit and system engineering applications.

Q110. Answer: B

Explanation: Coulomb's law describes electrostatic force magnitude as directly proportional to the product of charges and inversely proportional to the square of their separation. The medium's permittivity modifies the constant. Forces can be attractive or repulsive depending on charge signs. This law mirrors the inverse-square law of gravitation, forming a cornerstone of electrostatics and explaining atomic structure, field strength, and charge interactions in physics and electrical technology contexts.

Q111. Answer: A

Explanation: Magnetic flux equals the product of magnetic field strength, surface area, and cosine of the angle between field and normal vector. It measures total magnetic field passing through a surface, linking to Faraday's law of induction. Flux changes induce electromotive force in circuits. Understanding flux geometry helps design transformers, generators, and sensors. Alignment and surface orientation critically determine induced voltage and magnetic circuit efficiency in practice.

Q112. Answer: B

Explanation: Resonance happens when an external driving frequency matches a system's natural frequency, producing large amplitude oscillations. This energy transfer maximization underlies musical instruments, bridges, and electronic circuits. While resonance enhances performance, it can cause structural failure if uncontrolled. Understanding resonance enables safe and efficient design across mechanical, electrical, and acoustic systems, highlighting the balance between excitation frequency and inherent oscillatory properties in engineering and science contexts.

Q113. Answer: D

Explanation: Capacitance equals $\varepsilon A/d$; increasing plate area or decreasing separation enhances capacity to store charge per volt. Introducing a dielectric further increases capacitance by reducing effective electric field strength. This relationship enables design of capacitors for timing, filtering, and energy storage. Recognizing geometric and material factors allows engineers to optimize electronic performance and stability across varying temperature and voltage operating conditions in technological and industrial applications effectively.

Q114. Answer: B

Explanation: In SHM, acceleration is proportional to displacement from equilibrium and acts in the opposite direction ($a = -\omega^2 x$). This restoring acceleration drives oscillation, maintaining sinusoidal motion. At maximum displacement acceleration peaks; at equilibrium it is zero. This relationship defines oscillator dynamics in springs and pendulums, modeling vibrations in physics, engineering, and sound wave analysis across natural and mechanical oscillatory systems for accurate predictive analysis.

Q115. Answer: B

Explanation: Archimedes' principle states that a body submerged in a fluid experiences an upward buoyant force equal to the weight of the fluid displaced. This explains why objects float or sink depending on density comparison. The principle applies to liquids and gases, enabling ship design, fluid measurement, and density calculations. Understanding buoyancy integrates pressure and volume concepts essential in physics, hydrodynamics, and engineering system designs involving fluid equilibrium behavior effectively.

Q116. Answer: C

Explanation: Momentum conservation holds universally in isolated systems unaffected by external forces. Even when kinetic energy converts to heat or deformation (inelastic collisions), total vector momentum remains constant. This conservation law stems from Newton's third law symmetry. Applying momentum conservation simplifies solving collision and explosion problems in physics, engineering, and astrophysical events analysis, confirming physical system predictability and interaction consistency across scales and contexts of motion.

Q117. Answer: B

Explanation: Mechanical energy, the sum of potential and kinetic energies, stays constant if only conservative forces (like gravity and spring forces) act and no nonconservative forces (friction, air resistance) dissipate energy. This conservation principle simplifies motion prediction and underlies energy-based problem-solving. Recognizing conservative versus nonconservative interactions supports mechanical system design, efficiency evaluation, and energy management in physics, engineering, and environmental energy conservation contexts throughout practice.

Q118. Answer: C

Explanation: Adiabatic expansion involves no heat exchange (Q=0) with surroundings; internal energy decreases as gas does work, lowering temperature. This principle applies in thermodynamics, engines, and atmospheric processes. It contrasts with isothermal processes where temperature remains constant. Recognizing adiabatic behavior is crucial in designing efficient compressors, turbines, and understanding weather patterns where rapid expansion leads to cooling and condensation effects essential for system efficiency and natural science.

Q119. Answer: A

Explanation: Inside a conductor in electrostatic equilibrium, electric field equals zero because free charges move until internal fields cancel the applied one. As a result, excess charge resides on the surface. This principle explains shielding (Faraday cages) and conductor behavior in static conditions. Understanding electrostatic equilibrium ensures safe design of electronic enclosures and explains field distributions in materials, reinforcing fundamental electrostatics concepts in electrical engineering and physics.

Q120. Answer: B

Explanation: Frequency remains constant when waves pass from one medium to another because it depends on the source. Wave speed and wavelength change due to medium properties like density and elasticity. Understanding this invariance ensures proper explanation of refraction and Doppler phenomena. This principle links to energy transmission consistency, supporting analysis of sound, light, and seismic waves across interfaces in physics and engineering communication system design effectively.

Q121. Answer: C

Explanation: The magnetic field at the center of a circular loop equals $\mu_0 I/(2r)$, derived from Biot–Savart law, showing direct proportionality to current and inverse to radius. The field direction follows the right-hand rule around current flow. This relation aids in designing electromagnets, sensors, and MRI systems, linking current geometry and magnetic effects fundamental to electromagnetism and its numerous applications in science, medicine, and industry effectively.

Q122. Answer: B

Explanation: Nodes are points of zero displacement in standing waves formed by interference of identical waves traveling in opposite directions. Between nodes lie antinodes where displacement is maximum. Nodes indicate stationary regions due to destructive interference. This pattern occurs in musical instruments and resonant cavities. Understanding node-antinode structures clarifies resonance phenomena, frequency quantization, and energy distribution patterns in acoustics, optics, and quantum systems across scientific disciplines.

Q123. Answer: B

Explanation: In a fluid at rest, pressure acts equally in all directions at a point, a result of molecular motion and isotropy. This isotropic pressure explains hydraulic systems, buoyancy, and Pascal's law applications. Understanding uniform pressure distribution enables accurate prediction of forces on submerged surfaces and development of fluid-based machinery, confirming pressure's scalar nature and its essential role in hydrostatics, engineering, and atmospheric science contexts effectively.

Q124. Answer: C

Explanation: Interference between waves of close frequencies causes amplitude modulation called beats, perceived as periodic sound intensity variations. Beat frequency equals the difference between component frequencies. This concept aids instrument tuning, signal processing, and acoustic measurement. Understanding beats highlights constructive and destructive interference timing, illustrating wave superposition principles in sound, light, and electromagnetic applications used in engineering and music technology today.

Q125. Answer: C

Explanation: Lorentz force combines electric and magnetic effects on a moving charge, expressed as $F=q(E+v \times B)$. The electric term acts along the field, while the magnetic term acts perpendicularly to both velocity and field, altering direction but not speed. This equation governs charged particle motion in fields, foundational for cyclotrons, mass spectrometers, and plasma confinement systems central to modern physics and electrical engineering applications in analysis and design.

Q126. Answer: C

Explanation: Total momentum before and after an elastic collision is conserved. Initial momentum equals 5 kg × 4 m/s = 20 kg·m/s. No external forces act, so total momentum remains constant at 20 kg·m/s post-collision. While individual velocities change, the system's overall momentum is unchanged. This principle supports impact analysis, vehicle crash modeling, and kinetic interaction predictions across mechanical systems and particle collision scenarios in physics and applied engineering design processes.

Q127. Answer: B

Explanation: Efficiency of a heat engine equals work output divided by heat input, $\eta = W/Q_{in}$. Here, $\eta = 100/400 = 0.25$ or 25%. This ratio indicates how effectively the engine converts heat into mechanical work. The rest of the input energy dissipates as waste heat. Understanding efficiency helps improve thermal systems, engines, and energy conversion devices in engineering, ensuring optimal design for fuel economy and environmental sustainability.

Q128. Answer: B

Explanation: Hooke's law states that stress is proportional to strain within the elastic limit, $\sigma = E\varepsilon$. The constant of proportionality E is the Young's modulus, characterizing stiffness. Beyond this limit, deformation becomes plastic and irreversible. Linear behavior enables structural analysis, material testing, and spring design, ensuring elastic recovery in mechanical systems, components, and materials engineering where resilience and predictable response are required under mechanical stress.

Q129. Answer: C

Explanation: Photons are massless, chargeless particles that carry energy ($E = hf$) and momentum ($p = h/\lambda$). They travel at light speed in a vacuum and mediate electromagnetic interactions. Despite lacking rest mass, photons exert radiation pressure. Understanding photon properties bridges classical and quantum physics, explaining light behavior, radiation pressure, and photoelectric effects crucial to optics, astrophysics, and modern photonics technologies in applied scientific research and engineering.

Q130. Answer: B

Explanation: Escape velocity is derived from energy conservation, equating kinetic and gravitational potential energies: $v_e = \sqrt{2GM/R}$. It depends solely on planetary mass (M) and radius (R), not on the object's mass or atmosphere. This speed represents the minimum required to overcome gravitational attraction without propulsion. Understanding escape velocity informs rocket launch dynamics, planetary science, and celestial mechanics, linking energy conservation and gravitational field strength applications in astrophysics.

Q131. Answer: B

Explanation: The second law asserts that entropy, or disorder, in an isolated system tends to increase over time, making spontaneous processes irreversible. This principle governs energy directionality and efficiency limits. It explains why heat flows naturally from hot to cold, setting fundamental constraints for engines and refrigeration. Understanding entropy provides insight into natural processes, time asymmetry, and energy degradation across physics, chemistry, and environmental thermodynamics applications in analysis.

Q132. Answer: B

Explanation: The electron possesses the highest charge-to-mass ratio because of its small mass and single elementary charge. This property causes large deflection in magnetic and electric fields, as observed in cathode ray experiments that led to discovering the electron. Understanding charge-to-mass ratios is fundamental to particle physics, accelerator design, and determining subatomic particle properties crucial for quantum and electromagnetic interaction studies and applications across physics.

Q133. Answer: B

Explanation: The superposition principle applies to linear systems, stating that the resultant effect equals the vector sum of individual contributions. This holds for forces, fields, and waves where interactions are additive. Nonlinear systems violate this rule. Understanding superposition simplifies analyzing electric fields, mechanical vibrations, and wave interference, forming a core mathematical and conceptual foundation across mechanics, electromagnetism, and quantum mechanics for accurate modeling of complex physical systems.

Q134. Answer: B

Explanation: Wavelength equals wave speed divided by frequency ($\lambda = v/f$). At constant frequency, increasing speed increases wavelength. Sound travels faster in denser or more elastic media, so wavelength changes with medium while pitch (frequency) remains constant. Understanding wavelength relationships explains acoustics, sonar, and Doppler effects, enabling precise sound design, instrument tuning, and signal interpretation across various scientific, engineering, and communication technology applications worldwide.

Q135. Answer: A

Explanation: Conservation of angular momentum states that in the absence of external torque, the product of rotational inertia and angular velocity remains constant. When a skater pulls in arms, moment of inertia decreases, increasing spin rate to maintain constant angular momentum. This principle governs rotating systems from atomic to astronomical scales, explaining gyroscopic stability, planetary motion, and rotational energy management in mechanical engineering and physics systems design.

Q136. Answer: B

Explanation: Semiconductors show increased conductivity with temperature because heat excites more charge carriers across the band gap, reducing resistance. Unlike metals, their carrier density dominates conduction changes. This property enables temperature-sensitive devices, thermistors, and integrated circuits. Understanding semiconductor behavior underpins modern electronics, allowing engineers to tailor materials like silicon and germanium for optimal performance in transistors, sensors, and energy conversion technologies essential to digital systems today.

Q137. Answer: A

Explanation: Boyle's law states that for a fixed amount of gas at constant temperature, pressure is inversely proportional to volume (PV = constant). Doubling pressure halves volume, assuming ideal gas behavior. This relationship underlies pneumatic systems, lungs, and gas cylinder design. Understanding it aids in explaining atmospheric compression, vacuum physics, and energy storage in gases, forming part of the foundational ideal gas laws in thermodynamics and applied physics.

Q138. Answer: A

Explanation: The coulomb (C) is the SI unit of electric charge, representing the quantity of charge transported by one ampere of current in one second. This definition ties charge, current, and time directly. Understanding units of charge underlies electromagnetism, circuit theory, and electrochemistry, providing quantitative measures for electric interactions, current flow, and capacitor storage capacities in scientific and technological systems central to modern energy and electronics.

Q139. Answer: E

Explanation: Violet light has the shortest wavelength and highest frequency within the visible spectrum, giving it the highest photon energy (E=hf). Higher frequency increases energy per photon. This energy ordering explains phenomena like photoelectric emission thresholds and color separation in spectra. Understanding light energy relationships enhances optics, laser design, and photonic applications, connecting electromagnetic theory with practical energy quantization in visible radiation science and technology.

Q140. Answer: B

Explanation: Electric field strength E between parallel plates equals potential difference divided by separation, E = V/d. The field is uniform if edge effects are negligible, pointing from positive to negative plate. This relation defines capacitor behavior and field intensity estimation. Understanding this principle enables safe design of capacitors, electrostatic devices, and high-voltage equipment, linking field strength, potential, and geometry in electrostatics and electronic component engineering.

Q141. Answer: C

Explanation: When terminal velocity is reached, drag force and buoyant force balance weight, producing zero net force and constant velocity. This dynamic equilibrium demonstrates Newton's first law. Understanding it explains raindrop motion, parachuting, and sedimentation processes. Analyzing terminal velocity aids in modeling resistive motion, fluid dynamics, and industrial processes involving particle settling or aerodynamics within fluid environments across scientific and engineering disciplines.

Q142. Answer: B

Explanation: Hooke's law defines the spring restoring force as proportional and opposite to displacement ($F = -kx$). The negative sign indicates direction toward equilibrium. The spring constant k measures stiffness. This law models oscillations, vibrations, and mechanical energy storage. Understanding it allows precise control in mechanical systems, damping design, and energy transfer analysis, connecting microscopic elasticity to macroscopic engineering applications across various dynamic systems and instruments.

Q143. Answer: B

Explanation: Dispersion occurs because refractive index varies with wavelength; shorter wavelengths (violet) refract more than longer ones (red). This wavelength dependence separates white light into colors. Understanding dispersion explains rainbows, lens aberrations, and spectroscopic analysis. It forms the basis of optical design, enabling color correction, wavelength filtering, and light characterization in scientific, industrial, and artistic applications involving prisms and diffraction-based instruments for precise optical work.

Q144. Answer: C

Explanation: The henry (H) is the SI unit of inductance, defined as the induced electromotive force of one volt when current changes at one ampere per second. Inductance quantifies opposition to current change in coils. It's fundamental to electromagnetism, transformers, and energy storage in magnetic fields. Understanding inductance supports power electronics, wireless charging, and motor design, linking current variation, magnetic flux, and induced voltage across systems in physics applications.

Q145. Answer: A

Explanation: Refractive index $n = c/v$ represents the ratio of light's speed in vacuum to its speed in a material, determining light bending at boundaries. It varies with wavelength, explaining refraction and dispersion. This property underpins lens design, fiber optics, and optical instruments. Understanding refractive index enables precision in imaging, communications, and materials science, connecting wave propagation behavior with geometric optics and light-speed variations through media for analysis.

Q146. Answer: A

Explanation: Elastic potential energy in a spring equals $\frac{1}{2}kx^2$, derived from work done compressing or stretching by distance x against restoring force. Energy stored depends on spring constant and displacement squared. This concept explains oscillatory motion, mechanical energy conversion, and system resonance. Understanding potential energy relationships aids mechanical system design, vibration control, and safe material usage in automotive, aerospace, and industrial engineering for reliability and efficiency.

Q147. Answer: B

Explanation: Nuclear fission involves splitting heavy atomic nuclei into smaller fragments, releasing large amounts of energy due to mass-to-energy conversion ($E=mc^2$). This process powers nuclear reactors and atomic weapons. Understanding fission dynamics supports safe reactor control and energy generation, highlighting neutron moderation, chain reactions, and mass defect principles essential to nuclear physics, power technology, and radiological safety applications in energy science and engineering contexts.

Q148. Answer: C

Explanation: At resonance, inductive and capacitive reactances are equal and opposite, canceling each other, leaving only resistive impedance. The circuit current peaks and power factor reaches unity. This phenomenon is crucial in radio tuning, filters, and signal optimization. Understanding resonance conditions allows engineers to maximize power transfer and control frequencies in electronic and communication systems essential to practical signal processing and system efficiency.

Q149. Answer: C

Explanation: The combined gas law integrates Boyle's and Charles's laws, showing pressure-volume-temperature relationships for a fixed gas amount: $(P_1V_1)/T_1 = (P_2V_2)/T_2$. It expresses how gases respond to simultaneous changes in conditions. Understanding this law aids in weather prediction, engine analysis, and atmospheric science. It bridges to the ideal gas law, providing flexible frameworks for calculating state changes in physics, chemistry, and engineering thermodynamics problems effectively.

Q150. Answer: C

Explanation: A concave mirror forms a real, inverted image when the object is placed beyond the center of curvature. The image appears between focus and center, smaller than the object. Mirror geometry follows ray-tracing principles, confirming image formation rules for optical instruments. Understanding concave mirror behavior supports telescope, headlight, and imaging device design, connecting reflection, focal length, and geometric optics applications across physics and engineering visualization systems.

Q151. Answer: C

Explanation: The acceleration due to gravity varies slightly with latitude and altitude because Earth is not a perfect sphere—it bulges at the equator and rotates, causing centrifugal effects. Gravity is slightly stronger at the poles and weaker at the equator. Understanding these variations is important in geophysics, satellite navigation, and precision measurements, ensuring accurate data in engineering, mapping, and scientific applications involving gravitational effects worldwide.

Q152. Answer: C

Explanation: The kinetic theory of gases states that average molecular kinetic energy is proportional to absolute temperature ($E = 3/2\ kT$). Thus, temperature determines molecular motion intensity regardless of gas type or pressure. This microscopic interpretation connects temperature with energy, explaining pressure, diffusion, and heat transfer. Understanding this relationship underpins thermodynamics, meteorology, and materials science, describing gas behavior and molecular energy distribution in diverse physical systems.

Q153. Answer: C

Explanation: Evaporation occurs when molecules at a liquid's surface gain sufficient kinetic energy to overcome cohesive forces and enter the gas phase. This process cools the remaining liquid, forming the basis of sweating and natural cooling. Evaporation rate depends on temperature, surface area, and humidity. Understanding molecular escape behavior links thermodynamics, phase transitions, and energy balance in environmental physics, meteorology, and everyday physical processes such as drying and weather formation.

Q154. Answer: D

Explanation: Sound requires a medium to propagate because it transmits via particle vibrations. In a vacuum, there are no particles to carry the disturbance, so sound cannot travel. Understanding this limitation explains why outer space is silent and why communication requires electromagnetic waves there. It distinguishes mechanical and electromagnetic wave propagation, crucial for acoustics, engineering, and communication technologies across diverse media and scientific investigations in physics.

Q155. Answer: B

Explanation: The magnetic force on a current-carrying wire is given by $F = BIL \sin\theta$, where θ is the angle between current and magnetic field. When the wire is parallel ($\theta=0°$), $\sin\theta = 0$, yielding zero force. Maximum force occurs when perpendicular. This directional dependence guides electric motor and generator design, ensuring optimal torque generation and understanding magnetic interactions in applied physics and electromechanical systems effectively.

Q156. Answer: B

Explanation: Half-life represents the time for half the radioactive nuclei in a sample to decay into daughter products, a constant characteristic for each isotope. It follows exponential decay laws independent of initial quantity. Understanding half-life helps in carbon dating, medical imaging, and nuclear waste management. This concept connects nuclear physics with real-world time-dependent decay analysis used in energy, archaeology, and environmental monitoring across various scientific disciplines.

Q157. Answer: E

Explanation: Diamagnetism refers to a weak repulsion from magnetic fields caused by induced currents opposing the external field in atoms. All materials exhibit some diamagnetic effect, but in diamagnetic substances it dominates. Understanding diamagnetism explains magnetic levitation, stability, and materials' magnetic properties. It contrasts with paramagnetism and ferromagnetism, providing insights into electron motion and field interactions in materials science and applied magnetism.

Q158. Answer: E

Explanation: In a purely inductive circuit, voltage leads current by 90° because inductor opposition to current changes (back EMF) causes a quarter-cycle phase shift. This phase relation is fundamental to alternating current (AC) analysis and power factor correction. Understanding phase shifts helps optimize reactive circuit components and energy efficiency in power systems, transformers, and signal-processing devices across electrical engineering and applied physics.

Q159. Answer: C

Explanation: Buoyant force equals the weight of the fluid displaced, determined by fluid density and object's displaced volume ($F_b = \rho_{fluid} \cdot g \cdot V$). This is Archimedes' principle. Shape affects immersion depth but not total buoyant force. Understanding buoyancy explains flotation, ship design, and density measurements, highlighting equilibrium between gravitational and fluid forces across physics, engineering, and environmental systems for stability and structural efficiency.

Q160. Answer: C

Explanation: Total internal reflection occurs when light moves from a denser to a rarer medium at an angle exceeding the critical angle, causing all light to reflect internally. This phenomenon enables optical fiber transmission, diamond sparkle, and imaging devices. Understanding it connects refraction and reflection principles, critical-angle determination, and wave optics applications in telecommunications, fiber optics, and high-efficiency light transport systems across technology sectors.

Q161. Answer: B

Explanation: An ammeter measures electric current directly by being connected in series with the circuit so that current flows through it. It has low resistance to minimize circuit interference. The instrument's design, based on galvanometer principles, provides real-time current measurement in amperes. Understanding ammeter function supports safe electrical diagnostics, instrumentation calibration, and accurate analysis in electronics, power systems, and experimental physics practices involving current measurement.

Q162. Answer: B

Explanation: Magnetic field lines form circular, concentric loops around a straight current-carrying conductor, direction given by the right-hand rule. Field strength decreases with distance from the wire. Understanding this geometry is essential for designing coils, inductors, and electromagnets. It demonstrates Ampère's law and electromagnetic field behavior around conductors, central to electromagnetism theory and technological applications like power transmission, sensors, and magnetic-field mapping.

Q163. Answer: B

Explanation: Gravitational potential energy is defined as negative relative to infinity. As altitude increases, magnitude decreases (becomes less negative), reflecting weaker gravitational binding. The total mechanical energy of an orbiting satellite remains constant if frictionless. Understanding potential variation aids orbital mechanics, spacecraft design, and energy analysis in astrophysics, ensuring stability and efficient trajectory calculations for artificial satellites and celestial body interactions.

Q164. Answer: B

Explanation: The emitted light's color corresponds to photon energy differences between atomic energy levels ($E=hf$). Larger level separations produce higher-energy (shorter-wavelength) photons. This quantum property produces discrete emission spectra unique to each element. Understanding electronic transitions supports spectroscopy, chemical analysis, and quantum mechanics, linking atomic structure to observed light characteristics across laboratory diagnostics and astrophysical studies used in modern scientific exploration.

Q165. Answer: C

Explanation: The farad (F) measures capacitance, defined as one coulomb of charge stored per volt. It quantifies a capacitor's ability to store electrical energy in an electric field. Capacitance depends on plate area, separation, and dielectric properties. Understanding capacitance informs energy storage, filtering, and timing applications in circuits, highlighting electric field interactions and materials' dielectric behavior in electronics, physics, and power regulation technologies.

Q166. Answer: B

Explanation: Centripetal acceleration $a = v^2/r$ increases with the square of speed and decreases with radius. This inward acceleration maintains circular motion. Doubling speed quadruples required centripetal force. Understanding this proportionality helps design safe curves, centrifuges, and orbits. It highlights how motion parameters influence stability and required forces, fundamental for mechanical engineering, astrophysics, and physics of rotating systems across various technological and scientific contexts.

Q167. Answer: B

Explanation: In an ideal transformer, power is conserved: input power equals output power (P_primary = P_secondary), ignoring losses. Voltage and current adjust inversely with coil turns ratio. Understanding this ideal efficiency guides real transformer design, emphasizing minimal energy loss. It underpins electrical distribution systems, ensuring efficient voltage transformation, safe energy transfer, and sustainability across power transmission and electromechanical engineering technologies effectively.

Q168. Answer: B

Explanation: Frequency measures cycles per second with the SI unit hertz (Hz). It quantifies repetitive phenomena such as oscillations, waves, and alternating currents. Recognizing frequency's role connects time-domain and energy-domain analyses. It underpins AC power systems, signal processing, and sound perception, providing a fundamental descriptor for periodic motion, energy emission, and communication technologies across scientific, industrial, and engineering applications worldwide.

Q169. Answer: B

Explanation: A concave (diverging) lens always forms virtual, upright, and smaller images because light rays diverge after refraction, appearing to originate from a point on the same side of the lens. Such lenses correct myopia and serve in optical systems requiring field expansion. Understanding image formation enables precision in vision correction, optics design, and photography, connecting geometric optics to practical visualization and imaging system engineering effectively.

Q170. Answer: A

Explanation: As electrons drift through a conductor, collisions with atomic lattice ions cause resistive energy loss converted to heat, described by Joule heating ($P=I^2R$). This thermal effect limits current capacity and efficiency. Understanding energy dissipation informs material selection, cooling, and efficiency optimization in electrical systems. It connects atomic-scale interactions to macroscopic power loss, ensuring safe operation and energy conservation in engineering and applied physics contexts.

Q171. Answer: B

Explanation: Gravitational acceleration decreases slightly with altitude because distance from Earth's center increases, weakening gravitational force ($g \propto 1/r^2$). The effect is small within Earth's surface region but measurable in precise calculations. Understanding this variation informs satellite trajectories, ballistic computations, and engineering designs requiring precise gravitational modeling, linking classical gravitation theory to real-world applications across physics, aerospace, and geophysical sciences effectively.

Q172. Answer: B

Explanation: At the pendulum's mean position, potential energy is minimal while kinetic energy is maximal, giving maximum total mechanical energy exchange rate. At highest points, kinetic energy is zero, but total energy remains constant if no damping exists. Understanding energy distribution in pendulums clarifies oscillation dynamics and conservation laws. This concept applies to timekeeping, mechanical systems, and vibration analysis across physics, mechanics, and engineering technology design applications.

Q173. Answer: A

Explanation: The slope of a distance-time graph equals velocity, describing the rate of change of position. A straight line indicates constant speed; curved graphs show acceleration or deceleration. Graphical interpretation simplifies kinematic analysis and connects mathematical models with real motion. Understanding slope relationships aids visualization of motion parameters and reinforces analytical reasoning across mechanics, physics education, and engineering motion representation contexts in science and learning.

Q174. Answer: C

Explanation: In electromagnetic waves, energy oscillates equally between electric and magnetic fields, both contributing equally to total energy density. The fields are perpendicular to each other and to the direction of propagation. Understanding energy distribution aids in analyzing radiation intensity, antenna design, and wave propagation. This symmetry underlies electromagnetic theory's elegance and practical application in communication, optics, and energy transmission systems in physics and engineering.

Q175. Answer: B

Explanation: Specific heat capacity is the energy required to raise the temperature of 1 kg of material by 1°C (or 1 K). It measures a substance's ability to store heat. High specific heat moderates temperature changes, as seen in water's thermal stability. Understanding specific heat aids climate modeling, thermal design, and material analysis, connecting molecular energy storage to macroscopic temperature regulation in physics, engineering, and environmental applications.

Q176. Answer: B

Explanation: As sound propagates outward from its source, energy spreads over an expanding spherical surface, so intensity decreases inversely with the square of distance. This inverse-square law explains why distant sounds are quieter. Understanding this relationship aids in acoustics, noise control, and architectural design, ensuring proper sound distribution and attenuation in theaters, environmental studies, and communication systems in applied physics and engineering disciplines globally.

Q177. Answer: C

Explanation: Electric field lines originate on positive charges and terminate on negative charges, showing the direction of force on a positive test charge. The density of lines indicates field strength. Understanding field line patterns clarifies charge interactions, potential mapping, and electrostatic design principles for capacitors and shielding. This visualization aids comprehension of field behavior in physics, electronics, and electrostatic control across diverse applications and scientific analyses effectively.

Q178. Answer: B

Explanation: Upon entering a denser medium like glass, light's frequency remains constant, but speed decreases and wavelength shortens due to higher refractive index. This change causes refraction. Understanding this relationship underlies lens design, fiber optics, and optical phenomena. It connects electromagnetic wave theory with material interactions, ensuring precision in imaging, communications, and optical instrumentation across engineering and applied physics fields worldwide for accurate refraction control.

Q179. Answer: C

Explanation: A fuse protects circuits by melting and breaking the circuit when current exceeds safe limits, preventing overheating and fire. Made of low-melting-point metals, fuses act as safety devices in electrical systems. Understanding fuse function ensures safety and compliance in power distribution and electronic equipment. It illustrates current control, resistance heating, and energy protection concepts fundamental to circuit safety engineering and practical electrical design implementation.

Q180. Answer: B

Explanation: Torque on a dipole equals $pE \sin\theta$, where θ is the angle between dipole moment and electric field. Maximum torque occurs when $\theta = 90°$, i.e., the dipole is perpendicular. This torque aligns dipoles with the field, a concept vital to molecular physics, material polarization, and electric motor theory. Understanding torque interactions aids electrostatics, energy storage, and control of polar materials in physical and engineering systems.

Q181. Answer: C

Explanation: Transformers rely on magnetic flux coupling between primary and secondary coils through a shared magnetic core. Alternating current in the primary induces a changing magnetic field, generating voltage in the secondary by electromagnetic induction. Understanding magnetic coupling ensures efficient energy transfer and minimal losses. This principle supports electric power distribution, voltage conversion, and inductive charging systems across global electrical and electronic technologies.

Q182. Answer: A

Explanation: Kinetic energy of rotation equals $\frac{1}{2}I\omega^2$, analogous to translational energy ($\frac{1}{2}mv^2$), where I is moment of inertia and ω angular velocity. This energy depends on mass distribution about the axis. Understanding rotational energy enables analysis of turbines, wheels, and gyroscopes. It connects torque, inertia, and energy conservation, forming a cornerstone of rotational dynamics in physics, engineering, and applied mechanical design systems for efficiency.

Q183. Answer: B

Explanation: At constant temperature, gas pressure and volume are inversely proportional (Boyle's law: $PV = \text{constant}$). Reducing volume increases molecular collision frequency with container walls, raising pressure. This principle underlies pneumatic tools, lungs, and weather patterns. Understanding pressure-volume relationships explains compressibility and fluid mechanics in engineering and natural processes, emphasizing gas behavior consistency under thermodynamic constraints for practical mechanical and scientific systems.

Q184. Answer: A

Explanation: The elementary charge e equals 1.602×10^{-19} coulombs, a fundamental constant in physics. Electrons carry this charge negatively, while protons carry it positively. Recognizing this quantization of charge forms the basis for atomic structure, electric current, and electrochemical reactions. This value underpins electromagnetism, quantum physics, and circuit analysis, standardizing calculations across science, technology, and engineering disciplines involving charge-based interactions and electricity.

Q185. Answer: B

Explanation: Rayleigh scattering causes the blue color of the sky, as shorter wavelengths scatter more effectively by atmospheric molecules than longer red wavelengths. This wavelength-dependent scattering explains sunsets' red hue and sky's daytime color. Understanding scattering deepens knowledge of atmospheric optics, environmental monitoring, and astronomical observations, linking wave behavior to practical and natural visual phenomena across science, climate studies, and physics education.

Q186. Answer: C

Explanation: Planets orbit the Sun under gravitational attraction, balancing centripetal force with inertia. Newton's law of universal gravitation quantitatively describes this force. Understanding gravitational motion explains Kepler's laws, orbital mechanics, and tidal phenomena. This concept connects classical mechanics and astrophysics, forming the basis of celestial navigation, satellite technology, and understanding the solar system's dynamic equilibrium in scientific and engineering applications.

Q187. Answer: B

Explanation: Heating increases atomic vibration amplitude, slightly increasing average separation between molecules, causing expansion. This behavior underlies linear, area, and volume expansion relations in solids. Understanding expansion prevents structural stress in bridges and railways, enabling thermal compensation designs. It connects atomic motion to macroscopic changes, emphasizing temperature's effect on physical dimensions across materials science, thermodynamics, and engineering structural analysis for safety and efficiency.

Q188. Answer: D

Explanation: Hans Christian Oersted discovered in 1820 that an electric current deflects a nearby magnetic needle, revealing electricity's magnetic effect. This finding unified electricity and magnetism, inspiring electromagnetic theory. Understanding this discovery underpins electromagnet design, motors, and generators. It marks the origin of electrodynamics, bridging experimental physics and engineering, and shaping the development of modern electrical and communication technologies through applied scientific principles.

Q189. Answer: B

Explanation: The time period T of a wave is the reciprocal of its frequency ($T = 1/f$), representing time for one complete cycle. This relation connects frequency, wavelength, and velocity through $v = f\lambda$. Understanding period-frequency relationships underlies analysis of vibrations, oscillations, and signal processing, linking time and frequency domains crucial in communication, acoustics, and wave-based engineering across science and technology applications.

Q190. Answer: C

Explanation: Diffraction describes light's ability to bend around edges or through small openings, causing wave spreading and interference patterns. It demonstrates wave nature of light, crucial in understanding optical resolution and wave behavior. This phenomenon explains patterns in spectroscopy, imaging, and sound propagation. Understanding diffraction underpins modern optics, structural analysis, and photonics, connecting wave theory with technological applications in research and industrial imaging systems.

Q191. Answer: C

Explanation: Kinetic theory models gases as small particles moving randomly, colliding elastically without energy loss, and negligible intermolecular forces except during collisions. This assumption explains pressure, temperature, and volume behavior macroscopically. Understanding molecular motion supports thermodynamics, diffusion studies, and fluid dynamics. It bridges microscopic and macroscopic phenomena, forming a foundation for gas laws, transport processes, and modern physical chemistry across scientific disciplines.

Q192. Answer: A

Explanation: The electromagnetic spectrum organizes waves by increasing frequency or decreasing wavelength, from radio waves to gamma rays. This arrangement reflects energy progression ($E=hf$). Understanding the spectrum unites radio, microwave, infrared, visible, ultraviolet, X-ray, and gamma-ray regions. It underpins telecommunications, medical imaging, and astronomy, linking electromagnetic theory to practical technology across multiple scientific and engineering domains worldwide for effective analysis and innovation.

Q193. Answer: D

Explanation: Diamond possesses the highest refractive index among common materials (~2.42), meaning light slows most and bends strongly upon entering. This property contributes to its brilliance and dispersion. Understanding refractive indices aids in optical design, lens crafting, and gemology. It connects atomic structure to optical behavior, supporting applications in physics, materials science, and engineering optics, improving imaging precision and light management across technologies.

Q194. Answer: D

Explanation: Magnetic field inside a solenoid is $B = \mu_0 nI$; increasing current (I) or turns per unit length (n) strengthens the field. Inserting an iron core further amplifies it. This controllable field generation principle drives electromagnet design, motors, and inductors. Understanding solenoid field behavior integrates electromagnetism and material response, providing foundation for modern electrical, mechanical, and industrial engineering systems requiring adjustable magnetic forces.

Q195. Answer: B

Explanation: Floating occurs when an object's weight equals the buoyant force from displaced fluid, as per Archimedes' principle. The body's average density equals the fluid's density. This equilibrium determines floating stability and depth. Understanding buoyant balance aids shipbuilding, fluid design, and density measurement. It connects gravitational and pressure forces, forming key applications in marine physics, engineering, and environmental buoyancy-based technologies across fields.

Q196. Answer: A

Explanation: Electrical resistance varies directly with length and resistivity, inversely with cross-sectional area, and depends on temperature ($R=\rho L/A$). Longer conductors and smaller areas increase resistance. Understanding dependencies allows design optimization for electrical efficiency, heating, and safety. Material selection considers resistivity and thermal effects, essential in engineering wiring, power systems, and precision electronics across scientific and industrial electrical applications worldwide.

Q197. Answer: B

Explanation: Gauss's law states that total electric flux through a closed surface equals enclosed charge divided by permittivity of free space ($\Phi = q/\varepsilon_0$). The law holds regardless of surface shape, depending only on enclosed charge. Understanding Gauss's law simplifies electrostatic calculations, linking symmetry and charge distribution to electric fields, and underpins applications in capacitor design, shielding, and theoretical physics across science and electrical engineering disciplines.

Q198. Answer: A

Explanation: Gravitational potential energy equals mgh, representing energy stored due to position in a gravitational field. It converts to kinetic energy during descent, exemplifying energy conservation. Understanding potential energy aids system design, power estimation, and mechanical analysis in physics and engineering. It connects motion, height, and gravitational interaction principles across daily applications, mechanical devices, and scientific systems for energy transformation comprehension.

Q199. Answer: B

Explanation: In photoelectric effect, light intensity affects number of emitted electrons but not their individual energies, which depend on photon frequency. Higher intensity means more photons striking the surface per unit time, increasing emission count. Understanding this separation of intensity and frequency effects validated quantum theory, supporting modern photonics, sensors, and solar cell design in physics, electronics, and renewable energy technologies.

Q200. Answer: C

Explanation: Young's modulus measures material stiffness, given by the slope of stress-strain graph within the linear elastic region. Higher modulus indicates greater resistance to deformation. This mechanical property defines structural rigidity, guiding material selection in engineering. Understanding elastic moduli underpins construction, aerospace, and mechanical design, connecting atomic bonding strength with macroscopic mechanical behavior in science, technology, and industrial manufacturing applications globally.

Questions (201-300)

Q201. The capacitance of a parallel plate capacitor increases when:

A. Plate area decreases

B. Plate separation increases

C. Dielectric constant increases

D. Voltage decreases

E. Air is replaced with vacuum

Q202. The work done by a conservative force around a closed path is:

A. Maximum

B. Zero

C. Equal to potential energy

D. Negative always

E. Dependent on path length

Q203. The temperature at which the motion of gas molecules theoretically stops is:

A. 0°C

B. 100°C

C. –100°C

D. 0 K

E. –273°C

Q204. The energy of a photon is directly proportional to:

A. Wavelength

B. Frequency

C. Amplitude

D. Speed of light

E. Intensity

Q205. The coefficient of friction depends on:

A. Surface area of contact

B. Nature of surfaces in contact

C. Mass of object only

D. Speed of motion

E. Normal acceleration

Q206. The time rate of change of momentum of a body equals:

A. Impulse

B. Kinetic energy

C. Force

D. Work

E. Acceleration

Q207. The property of light that explains polarization is:

A. Particle nature

B. Wave nature

C. Reflection

D. Diffraction

E. Scattering

Q208. When a body moves with uniform circular motion, its:

A. Speed and velocity are constant

B. Speed constant but velocity changes

C. Velocity and acceleration constant

D. Acceleration zero
E. Kinetic energy constant and direction changes

Q209. The heating effect of an electric current is given by:
A. I/V
B. I^2R
C. V/I
D. IR
E. 1/IR

Q210. The process by which solids change directly into gas is called:
A. Condensation
B. Evaporation
C. Freezing
D. Sublimation
E. Deposition

Q211. When two waves of equal amplitude and frequency interfere destructively, the result is:
A. Increased amplitude
B. Beats formation
C. Random amplitude
D. Phase doubling
E. Zero amplitude

Q212. The process of converting alternating current into direct current is called:
A. Rectification
B. Amplification
C. Modulation
D. Induction
E. Polarization

Q213. The slope of a velocity-time graph represents:
A. Distance
B. Acceleration
C. Momentum
D. Force
E. Work

Q214. The resistance of a metallic wire increases with temperature because:
A. Electron density decreases
B. Atom vibrations increase
C. Electrons gain energy and move faster
D. Lattice expands reducing collisions
E. Potential difference reduces

Q215. The process by which heat transfers without actual movement of particles is:
A. Conduction
B. Convection
C. Radiation
D. Reflection
E. Diffusion

Q216. The magnetic field strength due to a long straight current-carrying wire varies:
A. Directly with distance
B. Inversely with distance
C. Directly with current squared

D. Inversely with current
E. Directly with resistance

Q217. The image formed by a plane mirror is:
A. Real, inverted
B. Virtual, upright, same size
C. Real, diminished
D. Virtual, inverted
E. Magnified and real

Q218. The unit of power in SI system is:
A. Joule
B. Watt
C. Newton
D. Ampere
E. Ohm

Q219. The color seen in thin oil films on water is due to:
A. Dispersion
B. Diffraction
C. Interference
D. Refraction
E. Scattering

Q220. The escape velocity on Earth is approximately:
A. 7.0 km/s
B. 8.0 km/s
C. 9.8 km/s
D. 11.2 km/s
E. 13.5 km/s

Q221. The rate of doing work is called:
A. Energy
B. Momentum
C. Power
D. Force
E. Efficiency

Q222. The phenomenon of bending of light when it passes through different media is:
A. Reflection
B. Refraction
C. Diffraction
D. Interference
E. Scattering

Q223. The quantity represented by the area under a force-displacement graph is:
A. Momentum
B. Energy rate
C. Power
D. Pressure
E. Work done

Q224. The device used to convert mechanical energy into electrical energy is:
A. Motor
B. Transformer
C. Generator

D. Battery
E. Rectifier

Q225. The process of energy emission from the surface of a hot body in the form of waves is called:
A. Convection
B. Radiation
C. Conduction
D. Reflection
E. Absorption

Q226. Which subatomic particle determines the identity of an element?
A. Neutron
B. Proton
C. Electron
D. Positron

Q227. What is the oxidation number of sulfur in H_2SO_4?
A. +4
B. +6
C. +2
D. 0

Q228. Which law states that equal volumes of gases at the same temperature and pressure contain equal numbers of molecules?
A. Boyle's Law
B. Charles's Law
C. Avogadro's Law
D. Dalton's Law

Q229. What type of bond forms between two nonmetal atoms sharing electrons?
A. Ionic bond
B. Metallic bond
C. Covalent bond
D. Hydrogen bond

Q230. Which process describes the direct conversion of a solid to a gas?
A. Deposition
B. Sublimation
C. Condensation
D. Evaporation

Q231. What is the molarity of a solution containing 2 moles of NaCl in 4 liters of solution?
A. 0.25 M
B. 0.5 M
C. 1 M
D. 2 M

Q232. Which statement about catalysts is true?
A. They alter the equilibrium position
B. They are consumed in reactions
C. They lower activation energy
D. They increase reaction enthalpy

Q233. Which orbital has the lowest energy in a hydrogen atom?
A. 2s
B. 1s
C. 3p

D. 2p

Q234. Which quantity is conserved in all chemical reactions?

A. Energy only

B. Mass only

C. Both mass and energy

D. Charge only

Q235. Which statement best defines an endothermic reaction?

A. It releases heat to surroundings

B. It absorbs heat from surroundings

C. It produces light only

D. It has a negative enthalpy change

Q236. Which principle explains the behavior of systems at equilibrium when disturbed?

A. Le Châtelier's Principle

B. Hess's Law

C. Dalton's Law

D. Boyle's Law

Q237. Which of the following compounds is polar?

A. CO_2

B. CH_4

C. H_2O

D. Cl_2

Q238. What is the pH of a 0.001 M HCl solution?

A. 1

B. 2

C. 3

D. 4

Q239. Which variable is held constant in an isothermal process?

A. Volume

B. Temperature

C. Pressure

D. Moles

Q240. Which of the following describes an ionic compound?

A. High melting point and conducts when molten

B. Soft and malleable

C. Low melting point and covalent bonding

D. Poor conductor in all states

Q241. Which term describes the minimum energy required for a reaction to occur?

A. Free energy

B. Activation energy

C. Potential energy

D. Reaction enthalpy

Q242. Which equation represents a balanced redox reaction?

A. $Zn + Cu^{2+} \rightarrow Zn^{2+} + Cu$

B. $Zn + Cu \rightarrow Zn + Cu^{2+}$

C. $Cu + Zn^{2+} \rightarrow Cu^{2+} + Zn$

D. $Zn^{2+} + Cu^{2+} \rightarrow Zn + Cu$

Q243. Which statement is true about reversible reactions?

A. They proceed only in one direction

B. They reach a dynamic equilibrium
C. They stop once products form
D. They violate energy conservation

Q244. What is the molar volume of an ideal gas at STP?
A. 22.4 L/mol
B. 1.0 L/mol
C. 11.2 L/mol
D. 24.5 L/mol

Q245. Which type of intermolecular force is strongest?
A. London dispersion
B. Dipole-dipole
C. Hydrogen bonding
D. Ionic bonding

Q246. What does the slope of a heating curve plateau represent?
A. Phase change at constant temperature
B. Increase in kinetic energy
C. Decrease in potential energy
D. Cooling process

Q247. What is the conjugate acid of NH_3?
A. NH_4^+
B. NH_2^-
C. H_2
D. OH^-

Q248. Which variable does the ideal gas constant (R) relate to in PV = nRT?
A. Pressure and temperature only
B. All state variables (P, V, n, T)
C. Volume and mass only
D. Energy and entropy

Q249. Which statement about metallic bonding is accurate?
A. Electrons are localized between atoms
B. It involves electron sharing between specific atoms
C. Delocalized electrons create conductivity and malleability
D. Bonds break easily under stress

Q250. What is the primary factor determining reaction rate in gases?
A. Pressure
B. Concentration of solid catalyst
C. Solvent polarity
D. Molecular collisions

Q251. What is the empirical formula of a compound containing 40% carbon, 6.7% hydrogen, and 53.3% oxygen?
A. CHO
B. CH_2O
C. $C_2H_4O_2$
D. $C_3H_6O_3$

Q252. Which factor increases the rate of a chemical reaction?
A. Decreasing temperature
B. Reducing surface area
C. Increasing concentration

D. Removing catalyst

Q253. What type of reaction is represented by $2KClO_3 \rightarrow 2KCl + 3O_2$?

A. Combination
B. Decomposition
C. Single replacement
D. Double replacement

Q254. Which term defines the heat absorbed or released during a reaction at constant pressure?

A. Entropy
B. Enthalpy
C. Gibbs free energy
D. Specific heat

Q255. What is the electron configuration of chlorine (atomic number 17)?

A. $1s^2 2s^2 2p^6 3s^2 3p^5$
B. $1s^2 2s^2 2p^6 3s^2 3p^6$
C. $1s^2 2s^2 2p^6 3s^1 3p^6$
D. $1s^2 2s^2 2p^6 3s^2 3p^4$

Q256. Which law describes the relationship between pressure and temperature at constant volume?

A. Boyle's Law
B. Charles's Law
C. Gay-Lussac's Law
D. Avogadro's Law

Q257. In an exothermic reaction, how do enthalpy and surroundings change?

A. $\Delta H > 0$; surroundings cool
B. $\Delta H < 0$; surroundings warm
C. $\Delta H > 0$; surroundings warm
D. $\Delta H = 0$; no change occurs

Q258. Which principle describes how electron orbitals fill in increasing energy order?

A. Pauli Exclusion Principle
B. Hund's Rule
C. Aufbau Principle
D. Dalton's Law

Q259. What occurs when a catalyst is added to a chemical reaction?

A. Reaction enthalpy changes
B. Activation energy decreases
C. Equilibrium constant increases
D. Product concentration doubles

Q260. Which property of water allows it to dissolve ionic substances effectively?

A. High molar mass
B. Nonpolar character
C. Polar nature and hydrogen bonding
D. Low specific heat

Q261. Which gas law relates pressure and volume inversely at constant temperature?

A. Charles's Law
B. Boyle's Law
C. Avogadro's Law
D. Dalton's Law

Q262. What does the equilibrium constant (K) express?

A. Reaction rate

B. Ratio of product to reactant concentrations at equilibrium
C. Initial concentration ratio
D. Energy released in reaction

Q263. What is the oxidation agent in the reaction $Zn + CuSO_4 \rightarrow ZnSO_4 + Cu$?
A. Cu^{2+}
B. Zn^{2+}
C. SO_4^{2-}
D. Zn

Q264. What determines the shape of a molecule in VSEPR theory?
A. Atomic masses
B. Number of electron pairs around central atom
C. Ionization energy
D. Bond enthalpy

Q265. What is the conjugate base of H_2SO_4?
A. SO_4^{2-}
B. HSO_4^-
C. OH^-
D. H_3O^+

Q266. Which element has the highest first ionization energy?
A. Sodium
B. Fluorine
C. Neon
D. Lithium

Q267. Which factor causes gas solubility in liquids to decrease?
A. Increasing pressure
B. Decreasing temperature
C. Increasing temperature
D. Adding solute

Q268. What is the pOH of a solution with $[OH^-] = 1 \times 10^{-3}$ M?
A. 3
B. 11
C. 10
D. 7

Q269. Which subatomic particle has negligible mass and a negative charge?
A. Proton
B. Neutron
C. Electron
D. Positron

Q270. Which process best explains why salt lowers the freezing point of water?
A. Colligative property effect
B. Hydrogen bond formation
C. Endothermic dissolution
D. Crystallization promotion

Q271. Which element is most electronegative?
A. Fluorine
B. Oxygen
C. Chlorine
D. Nitrogen

Q272. Which type of reaction combines two or more reactants into one product?
A. Decomposition
B. Combination (synthesis)
C. Single replacement
D. Double replacement
Q273. What is the oxidation number of chlorine in NaClO?
A. +1
B. -1
C. +3
D. +5
Q274. Which statement correctly describes endergonic reactions?
A. They release free energy spontaneously
B. They require energy input to proceed
C. They have negative ΔG
D. They occur without activation energy
Q275. Which type of solid conducts electricity when molten but not when solid?
A. Ionic solid
B. Covalent solid
C. Metallic solid
D. Molecular solid
Q276. Which quantum number defines the shape of an orbital?
A. Principal quantum number (n)
B. Azimuthal quantum number (l)
C. Magnetic quantum number (m)
D. Spin quantum number (s)
Q277. What is the mass of 2 moles of CO_2?
A. 22 g
B. 44 g
C. 88 g
D. 132 g
Q278. What is the pH of a 1×10^{-8} M HCl solution considering water's autoionization?
A. 8.00
B. 6.00
C. 7.00
D. 6.99
Q279. What is the primary reason ionic compounds have high melting points?
A. Weak molecular interactions
B. Strong electrostatic attraction between ions
C. Covalent bonding strength
D. Hydrogen bonding presence
Q280. What is the equilibrium constant expression for the reaction $N_2(g) + 3H_2(g) \rightleftharpoons 2NH_3(g)$?
A. $K = [N_2][H_2]^3/[NH_3]^2$
B. $K = [NH_3]^2/[N_2][H_2]^3$
C. $K = [N_2][H_2]/[NH_3]$
D. $K = [NH_3]/[H_2][N_2]^2$
Q281. Which term defines a reaction that can proceed in both directions under certain conditions?
A. Irreversible reaction
B. Reversible reaction

C. Spontaneous reaction
D. Catalyzed reaction

Q282. Which type of radiation has the greatest penetrating power?

A. Alpha
B. Beta
C. Gamma
D. Neutron

Q283. Which statement best describes a buffer solution?

A. It resists pH changes upon addition of small acid or base amounts
B. It neutralizes strong acids completely
C. It maintains a constant temperature
D. It acts only as a base

Q284. What is the oxidation state of chromium in Cr_2O_3?

A. +2
B. +3
C. +4
D. +6

Q285. What is the molar concentration of 0.5 mol of NaOH dissolved in 250 mL of water?

A. 1.0 M
B. 0.25 M
C. 2.0 M
D. 0.5 M

Q286. Which term represents the disorder or randomness in a system?

A. Enthalpy
B. Entropy
C. Free energy
D. Heat capacity

Q287. Which statement describes a saturated solution?

A. Contains less solute than can dissolve
B. Contains maximum solute at given conditions
C. Precipitates continuously
D. Cannot exist at high temperature

Q288. What is the effect of temperature increase on equilibrium in an exothermic reaction?

A. Shifts toward reactants
B. Shifts toward products
C. No effect
D. Increases rate only

Q289. Which species acts as the Brønsted-Lowry base in the reaction $NH_3 + H_2O \rightleftharpoons NH_4^+ + OH^-$?

A. NH_3
B. H_2O
C. NH_4^+
D. OH^-

Q290. What is the hybridization of carbon in methane (CH_4)?

A. sp
B. sp^2
C. sp^3
D. sp^3d

Q291. Which statement is true for an ideal gas?

A. Molecules attract each other
B. Collisions are inelastic
C. Volume of molecules is negligible
D. Internal energy depends on volume

Q292. What is the standard enthalpy change for elements in their standard states?

A. Zero
B. One
C. Variable
D. Negative

Q293. Which process describes oxidation?

A. Gain of electrons
B. Loss of electrons
C. Combination with hydrogen
D. Decrease in oxidation number

Q294. Which factor increases the solubility of solids in liquids?

A. Decreasing temperature
B. Increasing temperature
C. Lowering pressure
D. Removing solvent

Q295. What is the energy required to remove an electron from a gaseous atom called?

A. Electron affinity
B. Ionization energy
C. Lattice energy
D. Work function

Q296. Which gas law combines Boyle's, Charles's, and Gay-Lussac's laws?

A. Combined gas law
B. Dalton's law
C. Avogadro's law
D. Ideal gas law

Q297. Which metal is most reactive according to the activity series?

A. Copper
B. Zinc
C. Potassium
D. Silver

Q298. What does the term molar mass represent?

A. Number of atoms in one mole
B. Mass of one mole of a substance
C. Mass of one atom
D. Density of substance

Q299. What type of solid is diamond?

A. Ionic solid
B. Covalent network solid
C. Metallic solid
D. Molecular solid

Q300. Which factor affects the rate constant of a chemical reaction?

A. Reactant concentration
B. Temperature
C. Catalyst mass only

D. Product amount

Answer keys and Explanations (201-300)

Q201. Answer: C

Explanation: Capacitance $C = \varepsilon_0\varepsilon_r A/d$; it increases with larger dielectric constant ε_r, larger plate area A, or smaller distance d. Dielectric materials enhance charge storage without breakdown. Understanding capacitance control is essential for circuit design, filters, and energy storage. This principle connects material properties, geometry, and electric field behavior in electronic, physics, and engineering applications ensuring efficiency and high-performance capacitor systems across diverse technologies.

Q202. Answer: B

Explanation: For conservative forces like gravity or electrostatic force, total work done in a closed loop equals zero because potential energy depends only on initial and final positions. This property simplifies energy conservation analysis. Understanding conservative forces supports potential energy concepts, mechanical efficiency, and power system predictability in mechanics, physics, and engineering energy studies ensuring accurate work-energy evaluations in dynamic systems.

Q203. Answer: D

Explanation: Absolute zero (0 K or –273.15°C) marks the point where molecular motion ceases, defining the lower limit of the thermodynamic temperature scale. This concept underlies Kelvin temperature, cryogenics, and quantum physics. Understanding absolute zero clarifies kinetic energy behavior and matter's thermal properties, guiding research in superconductivity, low-temperature physics, and thermodynamic theory across science, technology, and applied engineering contexts globally.

Q204. Answer: B

Explanation: Photon energy equals Planck's constant times frequency ($E = hf$), showing direct proportionality. Higher frequency light carries more energy per photon, while wavelength inversely affects energy. This relationship explains photoelectric emission and quantum transitions. Understanding energy-frequency dependence connects electromagnetic theory and quantum physics, supporting technologies like lasers, spectroscopy, and photovoltaics across modern scientific and engineering applications worldwide effectively.

Q205. Answer: B

Explanation: The coefficient of friction is a dimensionless quantity depending on material interaction between surfaces, not contact area or mass. It reflects microscopic adhesion and roughness. Understanding friction coefficients aids in mechanical design, safety calculations, and wear reduction. It connects microscopic surface physics with macroscopic motion, ensuring efficiency and control in automotive, robotics, and industrial engineering systems globally through applied mechanics principles.

Q206. Answer: C

Explanation: Newton's second law defines force as the time derivative of momentum ($F = dp/dt$). It relates force to the rate of momentum change, forming a general definition applicable to variable mass and velocity systems. Understanding this relationship ensures dynamic equilibrium analysis and accurate motion predictions. It unifies classical mechanics, engineering design, and real-world applications such as collisions, propulsion, and structural force assessments in physics.

Q207. Answer: B

Explanation: Polarization occurs because light is a transverse wave, with vibrations perpendicular to propagation direction. It cannot arise in longitudinal waves. Understanding polarization supports optical filter design, stress analysis, and glare reduction. It connects electromagnetic wave theory with optical physics, enabling modern technologies like LCDs, photography, and communication systems, where controlling light's vibration plane is essential for clarity and precision in various scientific applications.

Q208. Answer: B

Explanation: In uniform circular motion, speed remains constant, but velocity continuously changes direction due to centripetal acceleration. This inward acceleration maintains circular path motion. Understanding velocity direction change clarifies rotational kinematics, centripetal force, and orbital dynamics. It connects Newtonian mechanics with real-world systems like satellites, turbines, and vehicle motion control across physics and engineering analysis for dynamic system stability understanding.

Q209. Answer: B

Explanation: The heating effect, or Joule heating, equals power dissipated as heat in a resistor ($P = I^2R$). Energy conversion from electrical to thermal form increases with current and resistance. Understanding this relationship ensures safety, material selection, and energy efficiency in devices. It connects electric current behavior with heat production principles in power electronics, circuit design, and practical applications of energy conversion in electrical systems.

Q210. Answer: D

Explanation: Sublimation is the direct transition from solid to gas without passing through liquid phase, as seen in dry ice and camphor. It occurs under low pressure and specific temperature conditions. Understanding sublimation connects phase transitions, molecular energy, and vapor pressure, vital in materials science, meteorology, and industrial processes like freeze-drying, purification, and crystal formation under controlled environments in applied physics and chemistry.

Q211. Answer: E

Explanation: Destructive interference occurs when two coherent waves meet 180° out of phase, causing amplitude cancellation ($A_{total} = 0$). This superposition results from wave phase opposition. Understanding interference aids in soundproofing, optical coatings, and communication technologies. It illustrates the wave nature of light and sound, providing core insight into diffraction, noise control, and interference pattern formation across applied physics and wave engineering disciplines.

Q212. Answer: A

Explanation: Rectification converts AC to DC using diodes, which allow current in one direction. Half-wave and full-wave circuits accomplish this electronically. Understanding rectification underpins power supplies, chargers, and electronic circuits. It connects semiconductor behavior to electrical conversion principles, ensuring stable DC for devices. This concept drives electronics design, renewable energy inverters, and electrical engineering systems for efficient energy utilization and conversion in practice.

Q213. Answer: B

Explanation: Velocity-time graph slope gives acceleration, the rate of velocity change. A constant slope indicates uniform acceleration. This graphical method simplifies motion analysis and relates directly to Newton's laws. Understanding slope interpretation connects mathematical modeling with real-world motion, enabling quantitative studies in kinematics, mechanics, and physics education for analyzing forces, acceleration, and speed variations effectively across engineering and scientific applications.

Q214. Answer: B

Explanation: As temperature rises, atomic vibrations intensify, increasing electron scattering and resistivity. This thermal dependence explains why metals conduct better at lower temperatures. Understanding resistance-temperature relations enables thermal management and sensor calibration. It connects microscopic electron behavior with macroscopic electrical properties, supporting precision circuit design, cryogenic studies, and temperature compensation techniques across electrical and materials engineering applications effectively.

Q215. Answer: C

Explanation: Radiation transfers heat via electromagnetic waves without medium requirement. Unlike conduction or convection, it operates in vacuum—e.g., solar heat reaching Earth. Understanding radiation informs thermal insulation, solar technology, and climate science. It connects thermodynamics with electromagnetic wave theory, supporting innovations in energy efficiency, spacecraft design, and environmental systems where direct energy transfer by radiation dominates thermal interactions globally.

Q216. Answer: B

Explanation: Magnetic field around a long straight conductor is $B = \mu_0 I/(2\pi r)$; it decreases inversely with distance r and increases with current I. This relation derives from Ampère's law. Understanding distance dependence supports magnetic field control, device calibration, and electromagnetic safety design. It connects fundamental electromagnetism with technological systems like power lines, inductors, and sensors in applied electrical engineering and physics research contexts.

Q217. Answer: B

Explanation: Plane mirrors produce virtual, upright images of same size as object, located behind mirror at equal distance. Light rays appear to diverge from virtual image point. Understanding image characteristics aids optical design, periscopes, and everyday reflections. It connects ray geometry and human perception principles, forming foundational concepts in optics, physics education, and engineering systems where accurate visual reproduction is crucial.

Q218. Answer: B

Explanation: The watt (W) equals one joule per second, measuring the rate of energy transfer or work done per unit time. Power quantifies energy consumption or output rate. Understanding power units enables energy efficiency analysis, engine performance evaluation, and circuit calculations. It bridges mechanical and electrical systems, forming a core parameter for physics, engineering, and applied energy systems analysis in global technology fields.

Q219. Answer: C

Explanation: Thin-film interference occurs when reflected light waves from film's top and bottom surfaces overlap, producing constructive or destructive interference depending on path difference. This effect generates colorful patterns in soap bubbles and oil films. Understanding interference enables precise measurement of thickness and optical coatings. It connects wave physics, optics, and color phenomena across engineering, art, and technology involving layered media systems.

Q220. Answer: D

Explanation: Earth's escape velocity equals $\sqrt{(2GM/R)}$, yielding about 11.2 km/s. It represents minimum speed needed to overcome gravitational attraction without propulsion. Understanding escape velocity informs rocket science, astrophysics, and planetary motion studies. It connects gravitational mechanics and energy conservation principles, supporting trajectory design and celestial mechanics in aerospace and space exploration engineering applications globally for precision and success.

Q221. Answer: C

Explanation: Power defines work done per unit time ($P = W/t$). It measures how quickly energy is transferred or mechanical tasks accomplished. This concept connects energy and time, influencing machine performance and efficiency analysis. Understanding power helps design engines, motors, and electrical systems, linking physics with engineering performance metrics and energy optimization across scientific and industrial fields effectively.

Q222. Answer: B

Explanation: Refraction occurs due to change in light speed when crossing media with different optical densities, causing direction bending. Snell's law quantifies this effect. Understanding refraction is key in lens design, optical fibers, and vision correction. It connects wave theory with geometry and material interaction, ensuring optical precision and control across physics, technology, and applied optical engineering for real-world implementations globally.

Q223. Answer: E

Explanation: The area under a force-displacement curve gives total work done ($W = \int F dx$). It represents energy transferred to or from an object through applied force. Understanding graphical interpretation of work aids mechanical analysis and energy studies. It connects kinematics, calculus, and physics principles, enhancing visualization and quantitative understanding in engineering, mechanics, and scientific analysis of motion and energy transfer.

Q224. Answer: C

Explanation: Generators employ electromagnetic induction to convert mechanical energy into electrical energy when a conductor moves through a magnetic field. This principle underlies all power generation. Understanding generation mechanisms ensures efficient electricity production, grid operation, and renewable integration. It connects motion, magnetism, and electricity, forming the cornerstone of electrical engineering, industrial design, and sustainable energy technologies worldwide through applied physics laws.

Q225. Answer: B

Explanation: Radiation involves energy emission via electromagnetic waves from any body with temperature above absolute zero. The Stefan–Boltzmann law quantifies emission proportional to fourth power of temperature. Understanding radiative transfer informs thermal control, astronomy, and energy balance systems. It links thermodynamics, wave physics, and material science, driving efficient heat management in engineering, space science, and environmental applications globally.

Q226. Answer: B

Explanation: The number of protons, known as the atomic number, defines the element and determines its position in the periodic table. Changing the proton count transforms one element into another, whereas altering neutrons forms isotopes and electrons influence charge. The proton count thus uniquely identifies each atom.

Q227. Answer: B

Explanation: Hydrogen has +1, and oxygen has -2 oxidation states. With two hydrogens and four oxygens, the total charge balance requires sulfur to be +6: $(2)(+1) + S + (4)(-2) = 0$. Therefore, sulfur's oxidation state is +6, consistent with its fully oxidized form in sulfuric acid.

Q228. Answer: C

Explanation: Avogadro's Law relates gas volume to the number of moles under constant temperature and pressure, asserting direct proportionality between volume and moles. This principle provides the foundation for the ideal gas law ($PV = nRT$), connecting molecular quantity to macroscopic gas behavior.

Q229. Answer: C

Explanation: Covalent bonds result from shared electron pairs between nonmetals, allowing each atom to achieve a stable electron configuration. Unlike ionic bonds, which involve electron transfer, covalent bonds create discrete molecules. Bond polarity depends on electronegativity differences between the bonded atoms.

Q230. Answer: B

Explanation: Sublimation occurs when a solid transitions directly into the gaseous phase without passing through a liquid state. Dry ice (solid CO_2) exemplifies this process. It requires energy input to overcome intermolecular forces, reflecting the endothermic nature of sublimation in phase-change thermodynamics.

Q231. Answer: B

Explanation: Molarity (M) equals moles of solute divided by liters of solution: $M = 2 \text{ mol} / 4 \text{ L} = 0.5 \text{ M}$. Molarity quantifies concentration and enables stoichiometric calculations in reactions, connecting solute amount to reaction yield and titration analysis.

Q232. Answer: C

Explanation: Catalysts speed up reactions by providing an alternative pathway with a lower activation energy barrier. They do not change equilibrium position or thermodynamic properties like enthalpy. Because catalysts remain unchanged after reactions, they can function repeatedly without being consumed.

Q233. Answer: B

Explanation: The 1s orbital is the closest to the nucleus and possesses the lowest energy level. According to the quantum mechanical model, electrons occupy orbitals from lowest to highest energy (Aufbau principle), starting with 1s, which forms the atom's ground-state configuration.

Q234. Answer: C

Explanation: According to the law of conservation of mass-energy, total mass and energy remain constant in a closed system. While chemical reactions convert energy forms, the overall balance remains. This principle underlies stoichiometry and thermodynamic equations like $\Delta E = q + w$.

Q235. Answer: B

Explanation: Endothermic reactions absorb energy, resulting in a positive enthalpy change ($\Delta H > 0$). This absorption causes a temperature drop in the surroundings. Examples include photosynthesis and thermal decomposition, where energy input is necessary to drive product formation.

Q236. Answer: A

Explanation: Le Châtelier's Principle states that a system at equilibrium adjusts to minimize the effect of any disturbance (concentration, temperature, or pressure changes). This self-regulation maintains dynamic equilibrium and predicts reaction shifts, essential in chemical process optimization.

Q237. Answer: C

Explanation: Water (H_2O) has a bent geometry and unequal electron distribution due to oxygen's higher electronegativity. This produces a dipole moment, making it polar. In contrast, CO_2 and CH_4 are nonpolar because their molecular symmetry cancels dipole effects.

Q238. Answer: C

Explanation: For strong acids like HCl, $[H^+]$ equals acid concentration. $pH = -\log[H^+] = -\log(0.001) = 3$. The logarithmic scale inversely relates to hydrogen ion concentration, meaning lower pH values correspond to stronger acidity in aqueous solutions.

Q239. Answer: B

Explanation: In an isothermal process, temperature remains constant, implying that internal energy (ΔU) does not change for ideal gases. Heat added is fully converted to work or vice versa, maintaining energy balance and reflecting thermodynamic principles of the first law.

Q240. Answer: A

Explanation: Ionic compounds consist of electrostatically bound ions, creating strong lattice structures with high melting points. They conduct electricity when molten or dissolved because ions are mobile, unlike covalent compounds, which typically have lower melting points and remain nonconductive.

Q241. Answer: B

Explanation: Activation energy is the minimum kinetic energy molecules need to collide effectively and form products. It determines reaction rate and can be lowered by catalysts, increasing successful collision frequency without altering overall reaction energetics.

Q242. Answer: A

Explanation: In this redox reaction, zinc is oxidized ($Zn \rightarrow Zn^{2+} + 2e^-$) while copper ions are reduced ($Cu^{2+} + 2e^- \rightarrow Cu$). The balanced form conserves mass and charge, showing electron transfer between metals, typical in galvanic cell reactions and electrochemical processes.

Q243. Answer: B

Explanation: Reversible reactions occur simultaneously in forward and reverse directions. At equilibrium, reaction rates equalize though molecular conversion continues dynamically. Concentrations remain constant, embodying chemical balance and illustrating dynamic equilibrium principles.

Q244. Answer: A

Explanation: At standard temperature (273 K) and pressure (1 atm), one mole of an ideal gas occupies 22.4 liters. This constant connects macroscopic volume to molar quantity and aids in stoichiometric conversions involving gaseous reactants and products.

Q245. Answer: C

Explanation: Hydrogen bonds occur when hydrogen is covalently bound to electronegative atoms like N, O, or F, creating a strong dipole attraction. These forces, though weaker than ionic or covalent bonds, are the strongest among intermolecular interactions and significantly affect boiling and melting points.

Q246. Answer: A

Explanation: A plateau in a heating curve indicates a phase change (melting or boiling) where temperature remains constant while energy breaks intermolecular bonds. During this period, added heat increases potential energy, not kinetic energy, reflecting latent heat absorption in phase transitions.

Q247. Answer: A

Explanation: Ammonia (NH_3) acts as a Bronsted-Lowry base by accepting a proton (H^+), forming ammonium (NH_4^+) as its conjugate acid. This relationship illustrates acid-base conjugate pairs differing by one proton, central to equilibrium understanding in aqueous acid-base systems.

Q248. Answer: B

Explanation: The universal gas constant (R) connects pressure, volume, moles, and temperature in the ideal gas law $PV = nRT$. It provides proportionality between macroscopic properties, enabling calculation of one variable when others are known under ideal conditions.

Q249. Answer: C

Explanation: In metallic bonds, delocalized electrons move freely among positive ion cores, enabling electrical conductivity, ductility, and malleability. The electron sea model explains these collective properties, distinguishing metals from ionic and covalent substances.

Q250. Answer: D

Explanation: In gaseous systems, reaction rate depends on effective molecular collisions. Frequency and energy of collisions influence product formation. Temperature and concentration changes alter kinetic energy and collision likelihood, aligning with the collision theory of chemical kinetics.

Q251. Answer: B

Explanation: To find the empirical formula, convert each percentage to moles by dividing by atomic masses: C (40/12 ≈ 3.33), H (6.7/1 ≈ 6.7), O (53.3/16 ≈ 3.33). Dividing by the smallest value (3.33) gives a mole ratio of 1:2:1, corresponding to CH_2O. This ratio represents the simplest whole-number formula showing the atomic composition.

Q252. Answer: C

Explanation: Increasing reactant concentration raises the number of particles per unit volume, leading to more frequent collisions. According to collision theory, higher collision frequency increases reaction rate if collisions have sufficient energy. Therefore, greater concentration accelerates chemical transformations, improving the probability of effective collisions forming products.

Q253. Answer: B

Explanation: This is a decomposition reaction in which a single compound, potassium chlorate, breaks down into potassium chloride and oxygen gas upon heating. Decomposition involves the breakdown of complex molecules into simpler products, often requiring energy input such as heat, electricity, or light to overcome chemical bond strengths.

Q254. Answer: B

Explanation: Enthalpy (ΔH) measures heat transfer in processes at constant pressure. A positive ΔH indicates endothermic reactions absorbing heat, while negative values represent exothermic reactions releasing heat. Enthalpy change reflects energy differences between reactants and products, governing thermal behavior in chemical transformations and calorimetric calculations.

Q255. Answer: A

Explanation: Chlorine, with 17 electrons, fills orbitals according to the Aufbau principle: $1s^2 2s^2 2p^6 3s^2 3p^5$. The outer $3p^5$ configuration reflects its halogen nature, requiring one electron to achieve noble gas stability. This arrangement determines chlorine's strong oxidizing behavior and its tendency to form -1 ions.

Q256. Answer: C

Explanation: Gay-Lussac's Law states that pressure is directly proportional to temperature when volume remains constant (P/T = constant). As temperature rises, gas molecules move faster, striking container walls with greater force, thereby increasing pressure. This law illustrates kinetic molecular theory's direct link between temperature and molecular motion.

Q257. Answer: B

Explanation: Exothermic reactions release heat, producing a negative enthalpy change ($\Delta H < 0$). Energy flows from the system to the surroundings, raising external temperature. Examples include combustion and neutralization reactions. This transfer demonstrates the conservation of energy and the interplay between system and environment in thermodynamics.

Q258. Answer: C

Explanation: The Aufbau Principle dictates that electrons occupy the lowest-energy orbitals first before moving to higher ones, creating stable configurations. This rule, combined with Hund's Rule and the Pauli Exclusion Principle, governs electron arrangements and explains periodic trends like valency and reactivity across elements.

Q259. Answer: B

Explanation: A catalyst lowers activation energy by providing an alternate pathway for reactants to reach products more efficiently. It does not affect thermodynamic quantities or equilibrium constant values but accelerates both forward and reverse reaction rates equally. Catalysts remain unchanged post-reaction, preserving overall system balance.

Q260. Answer: C

Explanation: Water's polarity and strong hydrogen bonds enable it to stabilize ions by surrounding them with opposite charges in solution, forming hydration shells. This electrostatic interaction weakens ionic bonds, leading to dissolution. Water's polar structure underpins its role as the universal solvent in chemistry and biology.

Q261. Answer: B

Explanation: Boyle's Law states that pressure is inversely proportional to volume when temperature and moles remain constant ($P_1V_1 = P_2V_2$). Decreasing volume increases collision frequency and pressure, consistent with kinetic molecular theory. This relationship underlies gas compression and expansion behavior in closed systems.

Q262. Answer: B

Explanation: The equilibrium constant quantifies the ratio of product and reactant concentrations at equilibrium for reversible reactions. A large K favors product formation, while a small K favors reactants. K remains constant at a given temperature, reflecting the balance between forward and reverse reaction tendencies.

Q263. Answer: A

Explanation: In this redox reaction, zinc is oxidized to Zn^{2+}, and copper ions (Cu^{2+}) are reduced to metallic copper. The oxidizing agent is the species reduced—in this case, Cu^{2+}—because it gains electrons. Identifying oxidizing and reducing agents clarifies electron transfer roles in electrochemical systems.

Q264. Answer: B

Explanation: Valence Shell Electron Pair Repulsion (VSEPR) theory predicts molecular geometry based on repulsion between electron pairs around the central atom. These pairs orient to minimize repulsion, determining shape. The model explains geometries such as linear, trigonal planar, and tetrahedral, correlating structure with bonding.

Q265. Answer: B

Explanation: When sulfuric acid donates one proton (H^+), it forms hydrogen sulfate (HSO_4^-) as its conjugate base. The conjugate base differs from the acid by one proton, maintaining charge balance. This acid-base pairing aligns with Bronsted-Lowry theory describing proton transfer equilibria in aqueous solutions.

Q266. Answer: C

Explanation: Neon, a noble gas, has a complete valence shell, making electron removal energetically difficult. Ionization energy increases across a period due to greater nuclear charge and decreases down a group as electron distance increases. Neon's stable configuration results in exceptionally high ionization energy.

Q267. Answer: C

Explanation: Gas solubility decreases as temperature increases because gas molecules gain kinetic energy, escaping the liquid phase more readily. This relationship is described by Henry's Law and observed in phenomena such as reduced oxygen content in warm water, impacting aquatic life and industrial applications.

Q268. Answer: A

Explanation: $pOH = -\log[OH^-] = -\log(1 \times 10^{-3}) = 3$. Since pH + pOH = 14 at 25°C, this solution's pH is 11, indicating basicity. pOH quantifies hydroxide ion concentration, complementing the pH scale for assessing acid-base balance in aqueous systems.

Q269. Answer: C

Explanation: Electrons carry a -1 charge and have negligible mass compared to protons or neutrons. Their mobility defines electrical conductivity and chemical bonding. The arrangement of electrons in energy levels determines reactivity, ion formation, and periodic behavior of elements across the table.

Q270. Answer: A

Explanation: Freezing point depression is a colligative property dependent on solute particle concentration, not identity. Dissolved salt ions disrupt ice lattice formation, requiring lower temperatures for solidification. This principle underlies road salting in winter and solution thermodynamics in phase equilibria.

Q271. Answer: A

Explanation: Fluorine, at the top right of the periodic table, exhibits the greatest tendency to attract shared electrons, defining maximum electronegativity. Its small atomic radius and high nuclear charge intensify electron pull. This property drives polarity in bonds and influences chemical reactivity patterns.

Q272. Answer: B

Explanation: Combination reactions merge two or more substances to form a single compound, typically releasing energy. For example, $2H_2 + O_2 \rightarrow 2H_2O$ illustrates synthesis. This type of reaction contrasts decomposition, which breaks compounds into simpler species under external energy input.

Q273. Answer: A

Explanation: In sodium hypochlorite, oxygen is -2 and sodium is +1. The sum must equal zero: (+1) + Cl + (-2) = 0, giving chlorine an oxidation state of +1. This compound serves as a mild oxidizing agent and disinfectant due to chlorine's intermediate oxidation state.

Q274. Answer: B

Explanation: Endergonic reactions possess positive Gibbs free energy ($\Delta G > 0$), requiring energy input to occur. They are nonspontaneous and often coupled with exergonic processes in biological or chemical systems to drive progress. Photosynthesis exemplifies endergonic energy conversion through light absorption.

Q275. Answer: A

Explanation: Ionic solids contain charged ions fixed in a rigid lattice that prevents electron movement when solid. Upon melting, ions become mobile, enabling electrical conduction. This distinction highlights ionic bonding's electrostatic nature and explains conductivity differences between solid and molten states.

Q276. Answer: B

Explanation: The azimuthal quantum number (l) specifies orbital shape and angular momentum. It distinguishes s, p, d, and f subshells, corresponding to values l = 0, 1, 2, and 3, respectively. This parameter influences orbital energy and geometry, shaping chemical bonding and electron cloud distribution patterns within atoms.

Q277. Answer: C

Explanation: The molar mass of CO_2 equals 12.01 g/mol (C) + 2 × 16.00 g/mol (O) = 44.01 g/mol. Therefore, 2 moles × 44.01 g/mol = 88 g. Molar mass links microscopic molecular structure to macroscopic quantities, allowing stoichiometric conversions between moles and grams in quantitative chemical analysis.

Q278. Answer: D

Explanation: When acid concentration is very low, water's self-ionization contributes significantly. Total $[H^+] \approx 1 \times 10^{-7} + 1 \times 10^{-8} = 1.1 \times 10^{-7}$. pH = $-\log(1.1 \times 10^{-7}) \approx 6.96$. The result is slightly acidic, illustrating the importance of autoionization effects at dilute acid concentrations.

Q279. Answer: B

Explanation: Ionic compounds exhibit strong electrostatic forces between oppositely charged ions arranged in crystal lattices. Overcoming these attractions requires substantial energy, leading to high melting and boiling points. The lattice energy magnitude correlates with ion charge and size, reflecting bond strength in ionic solids.

Q280. Answer: B

Explanation: For gas-phase equilibria, the equilibrium constant K is derived from the concentrations of products raised to their stoichiometric coefficients divided by reactants: $K = [NH_3]^2 / ([N_2][H_2]^3)$. This expression quantifies equilibrium composition, indicating whether a reaction favors product or reactant formation.

Q281. Answer: B

Explanation: Reversible reactions occur when products can revert to reactants until equilibrium is established. Both forward and reverse processes continue dynamically, maintaining constant concentrations. This behavior illustrates chemical balance and is fundamental in equilibrium systems like acid-base and gas-phase reactions.

Q282. Answer: C

Explanation: Gamma rays, composed of high-energy electromagnetic photons, possess no mass or charge, enabling deep penetration through materials. They require dense shielding like lead or concrete. In contrast, alpha and beta particles are easily stopped by paper or thin metal sheets due to lower energy.

Q283. Answer: A

Explanation: A buffer contains weak acid-base conjugate pairs that neutralize added acids or bases, minimizing pH fluctuations. For example, acetic acid and acetate maintain near-constant pH. Buffers are vital in biological systems and analytical chemistry to stabilize reaction environments against minor chemical disturbances.

Q284. Answer: B

Explanation: Oxygen contributes -2 each, giving a total of -6 for three oxygens. To balance neutrality, $2x + (-6) = 0$, so $x = +3$. Chromium thus has a +3 oxidation state, a common stable valence forming green oxides used as pigments and catalysts in chemical industries.

Q285. Answer: C

Explanation: Molarity (M) = moles/volume(L) = 0.5 mol / 0.25 L = 2.0 M. Molar concentration measures solute amount per liter, fundamental for quantitative chemical calculations. This measure aids in stoichiometry, titration, and reaction rate determinations in laboratory and industrial processes.

Q286. Answer: B

Explanation: Entropy (S) quantifies molecular disorder and energy dispersion. Greater entropy corresponds to more microstates and spontaneous tendencies toward equilibrium. Thermodynamic laws link entropy increase to spontaneous processes, driving natural phenomena such as diffusion, mixing, and thermal energy distribution.

Q287. Answer: B

Explanation: A saturated solution holds the maximum solute amount that dissolves at specific temperature and pressure. Additional solute remains undissolved, establishing equilibrium between dissolved and solid phases. Solubility depends on temperature, pressure, and solvent characteristics, influencing crystallization and precipitation processes.

Q288. Answer: A

Explanation: According to Le Châtelier's principle, adding heat to an exothermic reaction favors the reverse (endothermic) direction, shifting equilibrium toward reactants. This compensatory response reduces temperature impact and maintains balance, illustrating thermodynamic interplay between enthalpy and equilibrium state.

Q289. Answer: A

Explanation: Ammonia accepts a proton from water, forming NH_4^+ and OH^-. By definition, a Brønsted-Lowry base is a proton acceptor. This reaction produces a weakly basic solution, exemplifying equilibrium between conjugate acid-base pairs and illustrating reversible aqueous proton exchange mechanisms.

Q290. Answer: C

Explanation: Carbon in CH_4 forms four equivalent σ-bonds using sp^3 hybrid orbitals. The combination of one s and three p orbitals creates a tetrahedral structure with 109.5° bond angles. Hybridization explains molecular geometry and bonding strength within valence bond theory and molecular orbital frameworks.

Q291. Answer: C

Explanation: The ideal gas model assumes negligible molecular volume and no intermolecular forces, with collisions perfectly elastic. Internal energy depends solely on temperature. Though real gases deviate at high pressures or low temperatures, ideal gas behavior provides accurate approximations for many conditions.

Q292. Answer: A

Explanation: Elements in their standard states, such as O_2(g) or C(graphite), have $\Delta H^o = 0$ by definition. This convention establishes a reference baseline for calculating reaction enthalpies. It simplifies thermodynamic tables and enables consistent comparisons across chemical and physical transformations.

Q293. Answer: B

Explanation: Oxidation involves electron loss, increasing oxidation number. It commonly occurs alongside reduction in redox reactions, where one species donates electrons and another accepts them. Recognizing oxidation and reduction processes is fundamental to understanding electrochemistry and energy transfer mechanisms.

Q294. Answer: B

Explanation: Solubility of most solids rises with temperature because higher kinetic energy enhances molecular motion and solvation efficiency. Exceptions include exothermic dissolution processes. Temperature dependence governs recrystallization, purification, and industrial solubility control in chemical engineering operations.

Q295. Answer: B

Explanation: Ionization energy measures the minimum energy necessary to detach an electron from a gaseous atom or ion, forming a cation. It reflects atomic stability and electron-nucleus attraction. Ionization energy trends increase across periods and decrease down groups in the periodic table.

Q296. Answer: A

Explanation: The combined gas law unites three empirical laws: $(P_1V_1)/T_1 = (P_2V_2)/T_2$. It describes how pressure, volume, and temperature interrelate when gas quantity remains constant. This relationship allows predictions of gas behavior during simultaneous state variable changes under non-ideal conditions.

Q297. Answer: C

Explanation: Potassium, being highly electropositive, readily loses electrons to form K^+ ions. It reacts vigorously with water and oxygen, ranking near the top of the activity series. This high reactivity arises from low ionization energy and strong tendency to achieve noble gas configuration.

Q298. Answer: B

Explanation: Molar mass equals the mass of one mole (6.022×10^{23} particles) of a substance, expressed in grams per mole. It provides a bridge between atomic-scale and bulk quantities, essential for stoichiometric calculations and conversion between mass, moles, and volume in chemistry.

Q299. Answer: B

Explanation: Diamond consists of a continuous three-dimensional network of covalent bonds between carbon atoms, creating exceptional hardness and high melting point. The absence of free electrons explains its electrical insulating properties, distinguishing it from metallic and molecular solids structurally and functionally.

Q300. Answer: B

Explanation: The rate constant (k) varies with temperature according to the Arrhenius equation ($k = Ae^{-E_a/RT}$). Higher temperatures increase molecular kinetic energy, raising collision frequency and effective reaction rate. Catalysts influence rate by lowering activation energy, but only temperature directly alters k quantitatively.

Questions (301-400)

Q301. Which type of reaction is represented by NaOH + HCl → NaCl + H_2O?

A. Decomposition
B. Neutralization
C. Oxidation
D. Synthesis

Q302. Which instrument measures the heat of chemical reactions?

A. Calorimeter
B. Manometer
C. Spectrophotometer
D. Voltmeter

Q303. Which factor causes deviations from ideal gas behavior?

A. High temperature and low pressure
B. Strong intermolecular forces
C. Constant volume
D. Random molecular motion

Q304. What is the molar mass of calcium carbonate ($CaCO_3$)?

A. 60.01 g/mol
B. 100.09 g/mol
C. 120.11 g/mol
D. 84.05 g/mol

Q305. Which element forms a +2 ion by losing its outer electrons?

A. Sodium
B. Magnesium
C. Fluorine
D. Nitrogen

Q306. Which statement defines a redox reaction?

A. Reaction involving proton transfer
B. Exchange of ions between solutions
C. Simultaneous oxidation and reduction
D. Formation of precipitate

Q307. What happens to the rate of an endothermic reaction when temperature increases?

A. It decreases
B. It remains constant
C. It increases
D. It stops completely

Q308. Which type of radiation carries a +2 charge?

A. Alpha
B. Beta
C. Gamma
D. X-ray

Q309. What is the color of the flame in a sodium flame test?

A. Red
B. Green
C. Yellow
D. Blue

Q310. Which process involves loss of water from a compound?

A. Hydration
B. Hydrolysis
C. Dehydration
D. Neutralization

Q311. What is the pH of a neutral solution at 25°C?
A. 1
B. 7
C. 9
D. 14

Q312. Which factor determines boiling point elevation in solutions?
A. Solute particle concentration
B. Solvent density
C. Pressure alone
D. Temperature alone

Q313. What is the charge on the phosphate ion (PO_4)?
A. -1
B. -2
C. -3
D. +3

Q314. Which compound is an example of a weak acid?
A. HCl
B. HNO_3
C. HF
D. H_2SO_4

Q315. What does Hess's Law state?
A. Reaction enthalpy is independent of path taken
B. Reaction rate depends on activation energy
C. Volume and temperature vary inversely
D. Energy cannot be converted into work

Q316. Which orbital type holds a maximum of 10 electrons?
A. s
B. p
C. d
D. f

Q317. Which pair of elements is most likely to form an ionic compound?
A. Carbon and oxygen
B. Sodium and chlorine
C. Nitrogen and hydrogen
D. Sulfur and oxygen

Q318. Which gas law relates gas volume to temperature at constant pressure?
A. Boyle's Law
B. Graham's Law
C. Dalton's Law
D. Charles's Law

Q319. Which concept explains why reaction rates increase with temperature?
A. Increased collision frequency and energy
B. Decreased activation energy
C. Constant kinetic energy

D. Reduced molecular motion

Q320. What is the oxidation state of nitrogen in HNO_3?

A. +3

B. +5

C. +2

D. +4

Q321. Which property defines a strong electrolyte?

A. Poor conductor

B. Partial dissociation in water

C. Complete ionization producing high conductivity

D. Nonpolar molecular structure

Q322. What type of bond exists in Cl_2?

A. Polar covalent

B. Ionic

C. Nonpolar covalent

D. Metallic

Q323. What is the standard pressure used in gas law calculations?

A. 1 Pa

B. 760 mmHg

C. 273 K

D. 101.3°C

Q324. What is the oxidation number of oxygen in peroxides?

A. -1

B. -2

C. 0

D. +1

Q325. Which condition increases solubility of gases in liquids?

A. Higher temperature

B. Lower pressure

C. Higher pressure

D. Addition of solute

Q326. What happens to the pressure of a gas when its volume is halved at constant temperature?

A. Pressure doubles

B. Pressure halves

C. Pressure remains constant

D. Pressure decreases slightly

Q327. Which factor most influences the lattice energy of an ionic compound?

A. Atomic number

B. Ion size and charge

C. Temperature

D. Molar mass

Q328. What is the main product when calcium reacts with water?

A. Calcium oxide

B. Calcium hydroxide and hydrogen gas

C. Calcium carbonate

D. Calcium sulfate

Q329. Which term defines a homogeneous mixture of two or more substances?

A. Compound

B. Suspension
C. Colloid
D. Solution

Q330. What type of chemical bond involves the sharing of electron pairs between atoms?

A. Ionic bond
B. Covalent bond
C. Metallic bond
D. Hydrogen bond

Q331. Which of the following substances exhibits hydrogen bonding?

A. CH_4
B. H_2
C. NH_3
D. CO_2

Q332. What is the main assumption of the kinetic molecular theory?

A. Gases have attractive forces
B. Gas particles occupy significant volume
C. Gas particles move randomly with elastic collisions
D. Gas motion depends only on gravity

Q333. Which of these solutions has the highest boiling point?

A. 0.1 M NaCl
B. 0.1 M glucose
C. 0.2 M urea
D. 0.1 M $CaCl_2$

Q334. Which scientist proposed the plum pudding model of the atom?

A. Rutherford
B. Dalton
C. Thomson
D. Bohr

Q335. Which principle states that no two electrons can have the same set of quantum numbers?

A. Hund's Rule
B. Pauli Exclusion Principle
C. Aufbau Principle
D. Le Châtelier's Principle

Q336. Which property of metals allows them to conduct electricity?

A. Free movement of delocalized electrons
B. Strong ionic bonds
C. Weak van der Waals forces
D. Fixed electron positions

Q337. Which gas law states that volume is inversely proportional to pressure?

A. Boyle's Law
B. Charles's Law
C. Dalton's Law
D. Graham's Law

Q338. What is the oxidation number of sulfur in H_2SO_4?

A. +2
B. +4
C. +6
D. 0

Q339. Which process separates liquids based on boiling points?
A. Filtration
B. Distillation
C. Crystallization
D. Sublimation
Q340. Which element has the largest atomic radius in Period 2?
A. Lithium
B. Carbon
C. Oxygen
D. Fluorine
Q341. What is the formula for calculating percent composition?
A. (mass of solute / total mass) × 100
B. (moles of solute / volume) × 100
C. (moles of solvent / total moles) × 100
D. (mass of product / molar mass) × 100
Q342. Which type of reaction absorbs heat from surroundings?
A. Exothermic
B. Endothermic
C. Isothermal
D. Adiabatic
Q343. What is the name of the energy barrier reactants must overcome for a reaction to occur?
A. Activation energy
B. Enthalpy
C. Entropy
D. Potential energy
Q344. Which of these elements is a metalloid?
A. Sulfur
B. Silicon
C. Bromine
D. Calcium
Q345. Which process describes solid turning directly into gas?
A. Condensation
B. Vaporization
C. Sublimation
D. Deposition
Q346. What causes water's high specific heat capacity?
A. Ionic bonding
B. Hydrogen bonding
C. Covalent bonding
D. Metallic bonding
Q347. What happens to entropy when ice melts?
A. It decreases
B. It remains constant
C. It increases
D. It becomes zero
Q348. Which acid is present in vinegar?
A. Citric acid
B. Acetic acid

C. Hydrochloric acid
D. Formic acid
Q349. What is the product when iron reacts with oxygen in moist air?
A. $Fe_2O_3 \cdot xH_2O$ (rust)
B. Fe_3O_4
C. FeO
D. $Fe(OH)_2$
Q350. Which chemical law states that elements combine in fixed mass ratios?
A. Law of Multiple Proportions
B. Law of Definite Proportions
C. Law of Conservation of Energy
D. Law of Gaseous Volumes
Q351. What is the molecular geometry of carbon dioxide (CO_2)?
A. Bent
B. Linear
C. Trigonal planar
D. Tetrahedral
Q352. What happens when a strong acid reacts with a strong base?
A. Salt and water form with complete neutralization
B. Weak acid is produced
C. Gas evolution occurs
D. No reaction occurs
Q353. Which process separates a solid from a liquid using a porous barrier?
A. Evaporation
B. Crystallization
C. Distillation
D. Filtration
Q354. Which term describes the smallest unit of a compound retaining its chemical identity?
A. Atom
B. Molecule
C. Ion
D. Crystal
Q355. Which factor affects equilibrium according to Le Châtelier's principle?
A. Molecular weight
B. Changes in concentration, pressure, or temperature
C. Catalyst presence
D. Reaction container size
Q356. What type of reaction occurs when two ionic compounds exchange ions?
A. Synthesis
B. Decomposition
C. Double replacement
D. Combustion
Q357. What is the pH of a 1×10^{-4} M HCl solution?
A. 4
B. 10
C. 7
D. 3
Q358. What happens to the rate of a reaction when concentration of reactants increases?

A. It decreases
B. It remains constant
C. It increases
D. It stops

Q359. What type of intermolecular force is strongest?

A. Dispersion forces
B. Dipole-dipole interactions
C. Hydrogen bonding
D. London forces

Q360. Which process represents oxidation in redox reactions?

A. Gain of electrons
B. Loss of electrons
C. Formation of covalent bond
D. Proton transfer

Q361. Which element has the lowest electronegativity?

A. Fluorine
B. Cesium
C. Oxygen
D. Nitrogen

Q362. Which variable remains constant in isochoric processes?

A. Pressure
B. Temperature
C. Volume
D. Energy

Q363. What type of reaction is $2Na + Cl_2 \rightarrow 2NaCl$?

A. Decomposition
B. Synthesis
C. Combustion
D. Displacement

Q364. What is the oxidation number of nitrogen in NH_4^+?

A. -3
B. -2
C. -1
D. 0

Q365. Which gas diffuses fastest through a porous membrane?

A. Oxygen
B. Hydrogen
C. Nitrogen
D. Carbon dioxide

Q366. What is the role of enzymes in biochemical reactions?

A. Increase activation energy
B. Decrease reaction rate
C. Lower activation energy to speed reactions
D. Alter equilibrium position

Q367. Which term refers to the heat required to convert a liquid to vapor at constant temperature?

A. Enthalpy of vaporization
B. Enthalpy of fusion
C. Specific heat

D. Heat capacity

Q368. Which particle is responsible for chemical bonding?

A. Proton
B. Neutron
C. Valence electron
D. Photon

Q369. What is the conjugate acid of NH_3?

A. NH_2^-
B. NH_4^+
C. N_2H_4
D. NH

Q370. Which gas contributes most to the greenhouse effect?

A. Nitrogen
B. Oxygen
C. Carbon dioxide
D. Argon

Q371. What is the molarity of a solution containing 4 moles of solute in 2 liters of solution?

A. 1 M
B. 2 M
C. 3 M
D. 4 M

Q372. Which factor increases solubility of a gas in a liquid?

A. Increasing temperature
B. Increasing pressure
C. Decreasing pressure
D. Decreasing solvent polarity

Q373. What is the term for the energy change during bond formation?

A. Endothermic energy
B. Gibbs energy input
C. Entropy increase
D. Exothermic release

Q374. Which type of solution has equal osmotic pressure to that of a cell?

A. Isotonic
B. Hypertonic
C. Hypotonic
D. Supersaturated

Q375. What is the energy change when one mole of gaseous ions forms from atoms?

A. Electron affinity
B. Ionization energy
C. Lattice energy
D. Enthalpy of atomization

Q376. Which cellular organelle is primarily responsible for energy production through aerobic respiration?

A. Ribosome
B. Mitochondrion
C. Golgi apparatus
D. Lysosome

Q377. The primary function of ribosomes in a cell is to:

A. Store genetic material
B. Produce proteins
C. Generate ATP
D. Transport lipids

Q378. The phospholipid bilayer of the plasma membrane primarily provides:
A. Energy storage
B. Selective permeability
C. Protein synthesis
D. Genetic information storage

Q379. During photosynthesis, the light-dependent reactions occur in the:
A. Mitochondrial matrix
B. Cytoplasm
C. Thylakoid membranes
D. Nucleus

Q380. Which macromolecule serves as the main structural component of cell membranes?
A. Proteins
B. Lipids
C. Nucleic acids
D. Carbohydrates

Q381. The process by which cells maintain a stable internal environment is called:
A. Metabolism
B. Homeostasis
C. Osmosis
D. Respiration

Q382. Which structure controls the movement of substances into and out of the nucleus?
A. Nucleolus
B. Chromatin
C. Rough endoplasmic reticulum
D. Nuclear pore complex

Q383. The organelle responsible for modifying and packaging proteins for secretion is the:
A. Lysosome
B. Golgi apparatus
C. Nucleolus
D. Mitochondrion

Q384. In animal cells, which organelle is involved in digestion and recycling of cellular materials?
A. Ribosome
B. Lysosome
C. Centrosome
D. Chloroplast

Q385. The stage of mitosis when sister chromatids separate and move toward opposite poles is:
A. Prophase
B. Metaphase
C. Anaphase
D. Telophase

Q386. The enzyme responsible for synthesizing new DNA strands during replication is:
A. DNA polymerase
B. RNA polymerase
C. Ligase

D. Helicase

Q387. Which process converts glucose into pyruvate while generating small amounts of ATP?

A. Glycolysis
B. Krebs cycle
C. Electron transport chain
D. Fermentation

Q388. Which phase of the cell cycle involves DNA synthesis?

A. G1 phase
B. S phase
C. G2 phase
D. M phase

Q389. In plant cells, the organelle responsible for photosynthesis is the:

A. Chloroplast
B. Mitochondrion
C. Golgi apparatus
D. Lysosome

Q390. The site of lipid synthesis and detoxification in cells is the:

A. Rough endoplasmic reticulum
B. Golgi apparatus
C. Nucleus
D. Smooth endoplasmic reticulum

Q391. The fluid mosaic model describes:

A. Static lipid bilayer structure
B. Dynamic membrane composed of lipids and proteins
C. Random arrangement of carbohydrates
D. Fixed protein channels

Q392. Which of the following processes requires energy input by the cell?

A. Diffusion
B. Facilitated diffusion
C. Active transport
D. Osmosis

Q393. The mitotic spindle is composed primarily of:

A. Actin filaments
B. Microtubules
C. Intermediate filaments
D. Centrioles

Q394. The site where ribosomal RNA (rRNA) is synthesized is the:

A. Nucleolus
B. Nuclear envelope
C. Rough ER
D. Golgi apparatus

Q395. The structure responsible for separating chromosomes during mitosis is the:

A. Cell wall
B. Spindle apparatus
C. Golgi complex
D. Nuclear membrane

Q396. Enzymes act as catalysts by:

A. Increasing activation energy

B. Decreasing activation energy
C. Providing energy for reactions
D. Changing reaction equilibrium

Q397. Which phase of meiosis results in the separation of homologous chromosomes?

A. Prophase I
B. Metaphase I
C. Anaphase I
D. Telophase II

Q398. ATP synthase produces ATP by utilizing energy from:

A. Proton gradient
B. Electron flow through NADH
C. Oxygen binding
D. Carbon fixation

Q399. The cytoskeleton provides all of the following EXCEPT:

A. Cell shape
B. Intracellular transport
C. Protein synthesis
D. Structural support

Q400. The main purpose of the Calvin cycle in photosynthesis is to:

A. Generate ATP
B. Fix carbon dioxide into glucose
C. Produce oxygen
D. Capture light energy

Answer keys and Explanations (301-400)

Q301. Answer: B

Explanation: This reaction between a strong acid and strong base forms water and salt, classifying it as a neutralization reaction. Hydrogen ions from the acid combine with hydroxide ions from the base to produce water. Neutralization exemplifies double displacement processes, balancing ionic charges and demonstrating acid-base stoichiometry.

Q302. Answer: A

Explanation: A calorimeter quantifies heat exchange in physical or chemical processes by monitoring temperature changes. It enables calculation of enthalpy changes using $q = mC\Delta T$. Calorimetry provides experimental support for thermodynamic laws, evaluating heat transfer efficiency and energy balance in chemical systems.

Q303. Answer: B

Explanation: Real gases deviate from ideal gas predictions at high pressure or low temperature, where intermolecular forces become significant. These attractions reduce pressure and volume below ideal values. The van der Waals equation corrects for these deviations by including molecular size and interaction parameters.

Q304. Answer: B

Explanation: Molar mass = Ca (40.08) + C (12.01) + 3 × O (16.00) = 100.09 g/mol. This value represents one mole of calcium carbonate, a widely used compound in geological, biological, and industrial contexts, particularly in limestone, marble, and antacid formulations.

Q305. Answer: B

Explanation: Magnesium (atomic number 12) loses two 3s electrons to achieve the stable configuration of neon, forming Mg^{2+}. This behavior typifies Group 2 alkaline earth metals, which readily oxidize and participate in ionic bonding, producing stable ionic compounds such as MgO and $MgCl_2$.

Q306. Answer: C

Explanation: Redox reactions feature concurrent oxidation (electron loss) and reduction (electron gain) processes. Electron transfer couples species chemically, balancing charge and mass. Such reactions drive electrochemical cells, corrosion, metabolism, and combustion, reflecting fundamental energy exchange in matter transformations.

Q307. Answer: C

Explanation: Higher temperature raises kinetic energy of reactant molecules, enhancing collision frequency and energy. According to the Arrhenius equation, reaction rate increases exponentially with temperature, especially for endothermic processes requiring energy input to overcome activation barriers and initiate product formation.

Q308. Answer: A

Explanation: Alpha particles, composed of two protons and two neutrons, carry a +2 charge. Despite their high ionizing power, they have limited penetration due to mass and charge. They are emitted in radioactive decay of heavy nuclei such as uranium, altering atomic number and mass.

Q309. Answer: C

Explanation: Sodium atoms emit bright yellow light during flame excitation, resulting from electron transitions returning from excited to ground states. Each element's unique emission spectrum identifies it qualitatively, forming the basis of flame tests and atomic spectroscopy in analytical chemistry.

Q310. Answer: C

Explanation: Dehydration eliminates water molecules from a substance, often forming double bonds or anhydrides. It occurs in organic synthesis, acid catalysis, and thermal decomposition. For example, dehydration of alcohols produces alkenes, demonstrating chemical transformation via removal of H_2O under controlled conditions.

Q311. Answer: B

Explanation: In pure water at 25°C, $[H^+]$ equals $[OH^-] = 1 \times 10^{-7}$ M, yielding pH = 7. The neutral point reflects equal proton and hydroxide activity. Deviations indicate acidity or basicity, forming the foundation of pH scale interpretation for aqueous systems.

Q312. Answer: A

Explanation: Boiling point elevation is a colligative property depending on solute particle concentration, not identity. Nonvolatile solutes lower vapor pressure, requiring higher temperature for boiling. The relationship $\Delta T_\beta = iK_\beta m$ quantifies elevation, where i is van't Hoff factor, illustrating molecular effect on thermodynamic stability.

Q313. Answer: C

Explanation: Phosphate ion carries a -3 charge, resulting from oxygen atoms each contributing -2 oxidation state balanced by phosphorus at +5. This polyatomic ion is common in biological molecules like ATP and DNA and plays critical roles in energy transfer and biochemical buffering.

Q314. Answer: C

Explanation: Hydrofluoric acid (HF) only partially ionizes in water, categorizing it as a weak acid. Its strong hydrogen bonding resists complete dissociation, unlike HCl or HNO_3. Weak acids establish equilibrium between undissociated molecules and ions, influencing titration behavior and buffer formation.

Q315. Answer: A

Explanation: Hess's Law asserts that total enthalpy change equals the sum of enthalpy changes for individual steps, regardless of pathway. This principle allows indirect determination of reaction enthalpy through known intermediate processes, reinforcing the conservation of energy in thermochemistry.

Q316. Answer: C

Explanation: The d-subshell contains five orbitals, each accommodating two electrons with opposite spins, totaling ten. This capacity influences transition metal chemistry, explaining variable oxidation states and complex formation due to accessible d-orbitals enabling diverse bonding configurations.

Q317. Answer: B

Explanation: Ionic bonds form between metals and nonmetals through electron transfer. Sodium donates an electron to chlorine, producing Na^+ and Cl^- ions. Their strong electrostatic attraction yields sodium chloride, a crystalline solid with high melting point and electrical conductivity when molten or dissolved.

Q318. Answer: D

Explanation: Charles's Law states that the volume of a gas is directly proportional to its temperature in kelvin at constant pressure ($V_1/T_1 = V_2/T_2$). It demonstrates that increasing temperature causes expansion due to increased molecular kinetic energy, consistent with kinetic molecular theory.

Q319. Answer: A

Explanation: Temperature rise enhances molecular kinetic energy, increasing collision frequency and proportion of particles exceeding activation energy. According to collision theory and the Arrhenius model, these effects accelerate reactions exponentially, illustrating direct thermal influence on molecular dynamics.

Q320. Answer: B

Explanation: Hydrogen contributes +1, oxygen -2 × 3 = -6. Total charge must be zero: (+1) + N + (-6) = 0, so nitrogen = +5. This high oxidation state reflects nitric acid's strong oxidizing ability, common in nitration and industrial chemical processes.

Q321. Answer: C

Explanation: Strong electrolytes dissociate completely into ions in aqueous solution, ensuring high electrical conductivity. Examples include strong acids, bases, and soluble salts like NaCl. Their complete ionization distinguishes them from weak electrolytes, which partially dissociate, affecting current flow in electrochemical applications.

Q322. Answer: C

Explanation: The chlorine molecule (Cl_2) consists of identical atoms sharing electrons equally, forming a nonpolar covalent bond. The absence of electronegativity difference prevents dipole formation. Nonpolar bonds result in symmetric electron distribution and relatively low intermolecular forces compared to polar molecules.

Q323. Answer: B

Explanation: Standard pressure equals 1 atmosphere, equivalent to 760 mmHg or 101.3 kPa. This benchmark defines STP conditions, facilitating consistent gas law calculations and comparison of experimental data across varying laboratory environments worldwide.

Q324. Answer: A

Explanation: In peroxides such as H_2O_2, oxygen atoms share an O–O bond and exhibit -1 oxidation state, unlike the -2 state in most compounds. This reduced oxidation reflects partial bonding and contributes to peroxide's strong oxidizing reactivity in decomposition and redox processes.

Q325. Answer: C

Explanation: Gas solubility in liquids rises with increasing pressure, consistent with Henry's Law. Elevated pressure forces gas molecules into solution, enhancing concentration. This principle governs carbonation in beverages and gas absorption in industrial and environmental systems controlling dissolved gas behavior.

Q326. Answer: A

Explanation: According to Boyle's Law ($P_1V_1 = P_2V_2$), pressure and volume are inversely proportional at constant temperature. Halving the volume doubles collision frequency between gas molecules and container walls, doubling pressure. This relationship illustrates gas compressibility, fundamental in piston operation, scuba diving physics, and respiratory mechanisms.

Q327. Answer: B

Explanation: Lattice energy increases with greater ionic charge and decreases with larger ionic radius. Higher charges strengthen electrostatic attractions, requiring more energy to separate ions. Coulomb's law quantitatively relates lattice energy to charge and distance, explaining structural stability differences among ionic compounds.

Q328. Answer: B

Explanation: Calcium reacts moderately with water to yield calcium hydroxide and hydrogen gas: $Ca + 2H_2O \rightarrow Ca(OH)_2 + H_2$. This reaction demonstrates Group 2 metal reactivity trends, producing alkaline solutions due to hydroxide ions, typical of alkaline earth metal-water interactions.

Q329. Answer: D

Explanation: A solution features uniform distribution of solute particles within a solvent at molecular level, forming a single phase. Examples include saltwater and sugar solutions. Solutions differ from colloids or suspensions by lacking particle settling and scattering effects, ensuring consistent composition throughout.

Q330. Answer: B

Explanation: Covalent bonds form when atoms share valence electrons to achieve stability, commonly between nonmetals. Shared electron pairs create directional bonds influencing molecule shape. Covalent bonding explains molecular structure, polarity, and properties of organic and inorganic compounds essential in chemical reactivity.

Q331. Answer: C

Explanation: Ammonia (NH_3) forms hydrogen bonds due to highly electronegative nitrogen bonded to hydrogen. The resulting polarity and partial charges enable intermolecular attraction, influencing boiling points and solubility. Hydrogen bonding also stabilizes biological macromolecules such as DNA and proteins through weak yet significant forces.

Q332. Answer: C

Explanation: Kinetic molecular theory assumes gas molecules move randomly in straight lines, colliding elastically with no net energy loss. These collisions explain gas pressure and temperature relationships. The model idealizes behavior under low-pressure, high-temperature conditions, forming basis for gas law derivations.

Q333. Answer: D

Explanation: Boiling point elevation depends on number of solute particles. $CaCl_2$ dissociates into three ions (Ca^{2+} and $2Cl^-$), yielding the highest effective concentration. Colligative properties depend on total ion count, not identity, explaining stronger boiling point elevation for ionic compounds.

Q334. Answer: C

Explanation: J.J. Thomson's plum pudding model suggested electrons embedded in a positively charged matrix, resembling raisins in pudding. Though later disproved by Rutherford's nuclear model, Thomson's discovery of the electron revolutionized atomic theory and led to deeper understanding of subatomic structure.

Q335. Answer: B

Explanation: The Pauli Exclusion Principle asserts that each electron in an atom must have unique quantum numbers. Consequently, each orbital can contain a maximum of two electrons with opposite spins. This rule governs atomic structure, ensuring distinct electron configurations and element differentiation.

Q336. Answer: A

Explanation: Metallic bonding features delocalized valence electrons moving freely through a lattice of positive ions, enabling electrical and thermal conductivity. This electron mobility accounts for metallic luster, malleability, and ductility, distinguishing metals from covalent or ionic solids.

Q337. Answer: A

Explanation: Boyle's Law describes pressure-volume relationship: $P_1V_1 = P_2V_2$ at constant temperature. As pressure increases, volume decreases proportionally. This principle applies to breathing mechanics, hydraulic systems, and pneumatic devices relying on compressible gases for energy transfer.

Q338. Answer: C

Explanation: In H_2SO_4, hydrogen contributes +1 each and oxygen -2 each. The total charge equals zero: $2(+1) + S + 4(-2) = 0$, giving $S = +6$. Sulfur's high oxidation state explains sulfuric acid's strong oxidizing nature and wide industrial use in fertilizers and manufacturing.

Q339. Answer: B

Explanation: Distillation involves heating a liquid mixture to vaporize components sequentially by their distinct boiling points, then condensing vapors into separate fractions. It's essential for purifying solvents, refining petroleum, and isolating volatile substances from complex mixtures efficiently.

Q340. Answer: A

Explanation: Atomic radius decreases across a period due to increasing nuclear charge pulling electrons closer. Lithium, at the period's start, exhibits the largest radius. Understanding periodic trends explains reactivity, bond strength, and size-related behaviors among elements in the periodic table.

Q341. Answer: A

Explanation: Percent composition expresses the percentage of each component's mass relative to total compound mass. It helps determine empirical formulas and purity levels. By comparing constituent masses, chemists identify compound composition and analyze materials quantitatively in laboratory and industrial contexts.

Q342. Answer: B

Explanation: Endothermic reactions absorb heat, yielding positive enthalpy change ($\Delta H > 0$). Energy intake increases system potential, lowering external temperature. Examples include photosynthesis and thermal decomposition, where heat input breaks bonds and drives nonspontaneous transformations in chemical systems.

Q343. Answer: A

Explanation: Activation energy represents the minimum energy required for reactant molecules to reach the transition state and form products. Catalysts lower this barrier by providing alternate reaction pathways, increasing reaction rates without altering equilibrium positions or overall thermodynamics.

Q344. Answer: B

Explanation: Silicon exhibits both metallic and nonmetallic characteristics, classifying it as a metalloid. It conducts electricity moderately and forms covalent bonds, serving as a semiconductor in electronics. Metalloids bridge properties between metals and nonmetals across the periodic table.

Q345. Answer: C

Explanation: Sublimation occurs when solids vaporize without becoming liquid, as in dry ice (CO_2) or iodine crystals. This phase change involves overcoming intermolecular forces with sufficient energy, used industrially for purification and freeze-drying by direct solid-gas transition.

Q346. Answer: B

Explanation: Strong hydrogen bonds between water molecules require significant energy to increase temperature, giving water a high specific heat. This property stabilizes environmental and biological temperatures, reflecting hydrogen bonding's influence on molecular behavior and thermodynamic regulation.

Q347. Answer: C

Explanation: Melting transitions a solid's orderly structure into disordered liquid, increasing entropy. The rise in molecular randomness aligns with the second law of thermodynamics, which favors entropy increase in spontaneous processes such as melting, dissolution, or gas expansion.

Q348. Answer: B

Explanation: Vinegar contains acetic acid (CH_3COOH), a weak organic acid responsible for its sour taste and antimicrobial properties. It partially ionizes in water, forming acetate and hydronium ions, widely used in food preservation, cleaning, and laboratory titration.

Q349. Answer: A

Explanation: Iron oxidizes slowly in moist air forming hydrated ferric oxide, $Fe_2O_3 \cdot xH_2O$, known as rust. The process involves electrochemical oxidation of iron and reduction of oxygen, accelerated by water and electrolytes, causing material degradation over time.

Q350. Answer: B

Explanation: The Law of Definite Proportions asserts that a compound always contains elements in the same proportion by mass, regardless of sample size or source. This foundational law established chemical formula consistency, guiding stoichiometric relationships and molecular composition principles.

Q351. Answer: B

Explanation: Carbon dioxide has a linear geometry due to two regions of electron density around the central carbon atom, with no lone pairs. The double bonds to oxygen repel equally, resulting in a 180° bond angle. This geometry aligns with VSEPR theory, minimizing electron pair repulsion and explaining CO_2's nonpolar nature despite its polar bonds.

Q352. Answer: A

Explanation: Strong acid-base reactions, such as $HCl + NaOH \rightarrow NaCl + H_2O$, yield water and salt with nearly complete ionization. Both reactants dissociate fully, leading to neutral solutions with pH around 7. This complete neutralization illustrates stoichiometric proton transfer and ionic recombination fundamentals in aqueous chemistry.

Q353. Answer: D

Explanation: Filtration physically separates insoluble solids from liquids using a barrier, such as filter paper. The solid residue remains while the liquid filtrate passes through. It's widely employed in analytical and industrial chemistry to isolate precipitates or purify substances by removing particulate contaminants.

Q354. Answer: B

Explanation: A molecule consists of two or more atoms chemically bonded, representing the smallest entity that maintains compound properties. Covalent bonds hold these atoms together, and molecular behavior underpins reaction mechanisms, phase transitions, and physical characteristics of chemical substances.

Q355. Answer: B

Explanation: Le Châtelier's principle predicts that equilibrium shifts to counteract external changes in concentration, pressure, or temperature. Increasing reactant concentration drives the reaction forward, while heat addition favors endothermic direction. This principle helps regulate industrial reactions like the Haber process for ammonia synthesis.

Q356. Answer: C

Explanation: Double replacement reactions involve exchange of cations and anions between two ionic compounds, forming new products. For instance, $AgNO_3 + NaCl \rightarrow AgCl + NaNO_3$. Such reactions typically produce precipitates, gases, or water, exemplifying predictable ionic interactions in aqueous solutions.

Q357. Answer: A

Explanation: Since HCl dissociates completely, $[H^+] = 1 \times 10^{-4}$ M. pH = $-\log[H^+] = 4$, indicating acidity. The logarithmic pH scale quantifies proton concentration, facilitating comparison of acidic and basic strengths in aqueous solutions. Such calculations underpin acid-base titrations and equilibrium studies.

Q358. Answer: C

Explanation: Increasing reactant concentration raises particle frequency and collision probability, promoting more effective interactions per unit time. According to collision theory, higher concentration enhances rate until other limiting factors arise. This principle governs reaction kinetics in chemistry and industrial synthesis control.

Q359. Answer: C

Explanation: Hydrogen bonding, arising between hydrogen attached to N, O, or F and a lone pair on another electronegative atom, is the strongest intermolecular force. It significantly elevates boiling points and influences solubility. These interactions are crucial in water's properties and biological macromolecular stability.

Q360. Answer: B

Explanation: Oxidation is defined as electron loss, raising oxidation number. It often accompanies reduction of another species. For example, $Zn \rightarrow Zn^{2+} + 2e^-$ demonstrates electron loss. Identifying oxidation helps balance redox equations and predict direction in electrochemical and biological systems.

Q361. Answer: B

Explanation: Cesium, located at the lower left of the periodic table, exhibits the lowest electronegativity due to minimal effective nuclear charge on valence electrons. This weak electron attraction explains cesium's high reactivity and metallic character, common among alkali metals with low ionization energies.

Q362. Answer: C

Explanation: Isochoric (constant volume) processes occur without volume change, meaning no work is performed ($W = 0$). Energy transfers appear as heat, altering internal energy and pressure. Isochoric conditions apply to rigid containers, used in calorimetry and thermodynamic energy change measurements.

Q363. Answer: B

Explanation: This synthesis reaction combines sodium and chlorine to form sodium chloride. It exemplifies direct combination where reactants yield a single compound. Such reactions are energetically favorable, forming stable ionic lattices that release heat upon product formation.

Q364. Answer: A

Explanation: Hydrogen's oxidation state is +1; four hydrogens contribute +4. For NH_4^+, total charge is +1, so $N + 4(+1) = +1 \rightarrow N = -3$. This value aligns with nitrogen's tendency to form multiple oxidation states, highlighting its flexibility in redox and bonding chemistry.

Q365. Answer: B

Explanation: According to Graham's Law, diffusion rate is inversely proportional to the square root of molar mass. Hydrogen, being lightest, diffuses fastest. This property demonstrates kinetic molecular principles and underpins separation methods and leak detection in gas systems.

Q366. Answer: C

Explanation: Enzymes act as biological catalysts, providing alternative reaction pathways with reduced activation energy. They accelerate reactions without changing equilibrium positions. Specificity and efficiency make enzymes crucial in metabolism, cellular regulation, and industrial biocatalysis applications.

Q367. Answer: A

Explanation: Enthalpy of vaporization (ΔH_{vap}) represents energy absorbed to overcome intermolecular attractions during liquid-to-gas conversion. This endothermic process maintains temperature while phase changes occur, influencing boiling and condensation phenomena essential in distillation and energy transfer systems.

Q368. Answer: C

Explanation: Valence electrons, located in outermost shells, participate directly in chemical bonding. Their sharing, transfer, or delocalization defines covalent, ionic, and metallic bonds. Understanding valence interactions explains periodic trends, compound formation, and reactivity patterns across elements.

Q369. Answer: B

Explanation: When ammonia (NH_3) accepts a proton (H^+), it forms ammonium ion (NH_4^+). This pair demonstrates the Bronsted-Lowry concept of acids and bases, where conjugate species differ by one proton. Such equilibria are central to buffer systems and acid-base balance in aqueous chemistry.

Q370. Answer: C

Explanation: Carbon dioxide traps infrared radiation emitted by Earth, preventing heat loss and warming the atmosphere. Along with methane and water vapor, CO_2 is a key greenhouse gas. Its accumulation drives global climate change, motivating emission reduction and sustainable energy strategies.

Q371. Answer: B

Explanation: Molarity (M) = moles/volume(L). Substituting values: 4 mol / 2 L = 2 M. Molarity quantifies solute concentration, enabling stoichiometric calculations and standardization in titration, analytical chemistry, and chemical process control for accurate quantitative measurement.

Q372. Answer: B

Explanation: According to Henry's Law, gas solubility rises with pressure, as increased force compresses gas molecules into the liquid phase. This relationship governs carbonation in beverages and gas absorption in environmental systems, demonstrating pressure's impact on dissolution equilibria.

Q373. Answer: D

Explanation: Bond formation releases energy, making it an exothermic process. Energy release stabilizes products relative to reactants. Conversely, bond breaking absorbs energy. These opposing processes underpin thermochemistry, dictating whether reactions occur spontaneously and how energy flows between systems and surroundings.

Q374. Answer: A

Explanation: An isotonic solution exerts osmotic pressure equal to cellular fluids, preventing net water movement across membranes. Maintaining isotonicity is critical in medical and biological systems, ensuring cell stability during intravenous treatments and laboratory culture preservation.

Q375. Answer: B

Explanation: Ionization energy represents the energy required to remove an electron from an atom, forming a cation. It reflects atomic structure and effective nuclear charge. High ionization energy signifies strong electron binding, influencing reactivity trends and periodic properties across elements.

Q376. Answer: B

Explanation: Mitochondria generate ATP through oxidative phosphorylation, the final stage of aerobic respiration. They contain their own DNA and double membranes, reflecting evolutionary origins from endosymbiotic bacteria. The inner membrane's cristae provide surface area for electron transport chains. Understanding mitochondrial structure and function reveals energy conversion processes that sustain eukaryotic cells and highlights organelle specialization vital to metabolic efficiency and cellular respiration in multicellular organisms.

Q377. Answer: B

Explanation: Ribosomes are molecular machines that synthesize proteins by translating messenger RNA sequences into polypeptide chains. Found free in cytoplasm or bound to the rough endoplasmic reticulum, ribosomes are composed of rRNA and proteins. Their function links transcription and translation, ensuring genetic information becomes cellular structure and enzymes. Understanding ribosome activity underpins knowledge of gene expression, growth, and responses in prokaryotic and eukaryotic cells.

Q378. Answer: B

Explanation: The phospholipid bilayer forms the structural basis of cell membranes, offering selective permeability that regulates molecular movement. Hydrophobic tails face inward while hydrophilic heads face aqueous environments. Embedded proteins facilitate transport and signaling. This dynamic arrangement maintains internal homeostasis and communication. Understanding membrane structure explains diffusion, osmosis, and transport mechanisms essential for cellular survival, energy regulation, and environmental adaptation across biological systems.

Q379. Answer: C

Explanation: Light-dependent reactions occur within chloroplast thylakoid membranes, where chlorophyll captures light energy to produce ATP and NADPH. These molecules power the Calvin cycle in the stroma. Electron transport and photolysis of water release oxygen. Understanding this process connects solar energy conversion to carbon fixation, fundamental to life's energy flow and ecosystem productivity through biochemical reactions of photosynthesis in plant and algal cells.

Q380. Answer: B

Explanation: Lipids, primarily phospholipids, form the core structure of biological membranes, creating hydrophobic barriers between internal and external environments. Cholesterol and glycolipids provide fluidity and stability. Proteins embedded within lipid layers perform signaling and transport. Understanding lipid organization explains how membranes maintain integrity, regulate transport, and enable communication, making them vital for cell compartmentalization, function, and adaptation in all living organisms across domains.

Q381. Answer: B

Explanation: Homeostasis is the regulation of internal conditions to sustain optimal function despite external fluctuations. It involves feedback mechanisms controlling temperature, pH, and ion concentration. Cellular membranes, enzymes, and signaling networks coordinate responses to change. Understanding homeostasis integrates physiology and molecular biology, emphasizing how cellular systems sustain balance essential for survival, growth, and efficient metabolic regulation within living organisms across environments.

Q382. Answer: D

Explanation: Nuclear pore complexes regulate molecular traffic between cytoplasm and nucleus, allowing selective exchange of RNA, proteins, and signaling molecules. These multiprotein structures maintain genomic integrity while coordinating transcriptional regulation. Understanding nuclear transport mechanisms clarifies gene expression control and compartmentalization in eukaryotic cells, linking structural biology with cellular regulation processes vital for maintaining DNA stability and coordinated molecular communication within the cell.

Q383. Answer: B

Explanation: The Golgi apparatus modifies, sorts, and packages proteins synthesized in the rough ER for secretion or intracellular use. Through cisternae stacks, it adds carbohydrate or phosphate groups forming glycoproteins. Vesicles transport products to final destinations. Understanding Golgi function clarifies post-translational processing and cellular logistics essential to secretion, membrane maintenance, and communication across eukaryotic cells ensuring efficient protein trafficking and metabolic coordination.

Q384. Answer: B

Explanation: Lysosomes contain hydrolytic enzymes that degrade macromolecules, damaged organelles, and pathogens. Formed by the Golgi apparatus, they maintain acidic conditions for enzyme activity. Understanding lysosomal digestion explains waste management, apoptosis, and cellular renewal processes. Malfunction leads to storage diseases. These recycling functions underscore cellular maintenance and defense, vital for tissue health and energy recovery across multicellular organisms and molecular biology systems.

Q385. Answer: C

Explanation: During anaphase, spindle fibers shorten, pulling sister chromatids apart toward opposite poles ensuring equal genetic distribution. This event follows alignment in metaphase and precedes cytokinesis. Understanding mitotic stages provides insight into chromosome segregation, genetic stability, and cell cycle control. Accurate chromatid separation prevents mutations, ensuring healthy growth, development, and replication fidelity essential to multicellular organism function and genetic integrity.

Q386. Answer: A

Explanation: DNA polymerase synthesizes new complementary strands by adding nucleotides to a template using base-pairing rules. It ensures replication accuracy through proofreading. This process is central to genetic inheritance and cellular division. Understanding DNA polymerase function clarifies replication mechanics, mutation prevention, and biotechnological applications, reinforcing molecular biology's foundational concepts in heredity, evolution, and cell cycle regulation critical for living systems stability.

Q387. Answer: A

Explanation: Glycolysis occurs in the cytoplasm, breaking glucose into two pyruvate molecules and producing a net gain of two ATP via substrate-level phosphorylation. It functions anaerobically and serves as the first step in cellular respiration. Understanding glycolysis links carbohydrate metabolism, energy extraction, and metabolic regulation, illustrating how organisms harness chemical energy foundational to cellular survival and energy flow through biological systems.

Q388. Answer: B

Explanation: The S phase (synthesis phase) of the cell cycle is when DNA replication occurs, doubling genetic content before mitosis. Accurate duplication ensures genetic consistency. Understanding S phase control clarifies cell cycle checkpoints and replication fidelity, key in studying cancer biology and cell growth. Regulation during S phase underlines molecular mechanisms maintaining genome stability across eukaryotic life forms through precise DNA synthesis coordination and monitoring.

Q389. Answer: A

Explanation: Chloroplasts capture sunlight to drive photosynthesis, converting carbon dioxide and water into glucose and oxygen. Containing chlorophyll pigments and thylakoid membranes, chloroplasts transform radiant energy into chemical energy stored in ATP and sugars. Understanding chloroplast function connects cellular energetics, ecology, and evolution, revealing how autotrophic organisms sustain food chains and planetary energy cycles through photosynthetic biochemical processes in biological systems.

Q390. Answer: D

Explanation: The smooth endoplasmic reticulum synthesizes lipids, phospholipids, and steroids while detoxifying harmful substances, especially in liver cells. Lacking ribosomes, it complements rough ER's protein synthesis. Understanding smooth ER function reveals cellular adaptation to metabolic needs, drug detoxification, and hormone production. It integrates biochemistry, physiology, and molecular structure, showcasing organelle specialization and essential metabolic versatility supporting organismal health and homeostasis.

Q391. Answer: B

Explanation: The fluid mosaic model depicts membranes as flexible bilayers of phospholipids embedded with proteins that move laterally. This dynamic arrangement supports transport, signaling, and enzymatic functions. Cholesterol modulates fluidity, while glycoproteins aid recognition. Understanding this model integrates biophysics and cellular biology, explaining membrane behavior critical for communication, regulation, and compartmentalization fundamental to cellular integrity and adaptability across living organisms.

Q392. Answer: C

Explanation: Active transport moves molecules against concentration gradients using ATP or electrochemical energy via membrane pumps. This process maintains ion balances and nutrient uptake essential for cellular function. Understanding active transport elucidates mechanisms of nerve signaling, muscle contraction, and metabolic regulation. It distinguishes energy-dependent movement from passive diffusion, underscoring cellular energetics central to physiology and molecular biology in living systems.

Q393. Answer: B

Explanation: The mitotic spindle consists mainly of microtubules, dynamic polymers that attach to kinetochores to separate chromosomes during cell division. Microtubules, formed from tubulin, orchestrate chromosome alignment and segregation. Understanding spindle dynamics provides insight into mitosis regulation, drug targeting (e.g., taxol), and cytoskeletal organization. It links structural biology with cellular mechanics, ensuring genetic stability across eukaryotic organisms during replication and development processes.

Q394. Answer: A

Explanation: The nucleolus, a dense nuclear region, synthesizes and assembles ribosomal RNA and ribosomal subunits before export to the cytoplasm. It supports protein synthesis by ensuring ribosome availability. Understanding nucleolar function connects gene expression, RNA processing, and cellular growth regulation. It highlights the interplay between transcriptional activity and protein production, essential for cellular metabolism and overall organismal development in eukaryotic biology systems.

Q395. Answer: B

Explanation: The spindle apparatus, composed of microtubules and associated proteins, ensures chromosome movement and segregation during mitosis. Anchored by centrosomes, it attaches to kinetochores on chromosomes. Understanding spindle function provides insight into mitotic accuracy, cancer biology, and pharmacological control of cell division. It exemplifies cytoskeletal coordination essential for maintaining genetic integrity and organized cell replication in eukaryotic systems during growth and repair.

Q396. Answer: B

Explanation: Enzymes lower the activation energy required for chemical reactions, enabling biological processes to occur efficiently at physiological temperatures. They bind substrates at active sites, forming enzyme-substrate complexes that stabilize transition states. Understanding enzymatic catalysis reveals molecular control mechanisms, metabolic pathways, and homeostatic regulation central to life processes. It unites chemistry and biology in explaining energy transformation and biochemical specificity in living systems.

Q397. Answer: C

Explanation: During anaphase I, homologous chromosomes are pulled toward opposite poles, halving chromosome number and ensuring genetic diversity through recombination. Sister chromatids remain joined until meiosis II. Understanding this reduction division explains gametogenesis, inheritance patterns, and variation. It integrates genetics and cytology, highlighting meiosis as a key evolutionary process underpinning reproduction, diversity, and chromosomal behavior in sexually reproducing organisms throughout biology.

Q398. Answer: A

Explanation: ATP synthase synthesizes ATP as protons move down their electrochemical gradient across the inner mitochondrial membrane. This chemiosmotic process couples proton motive force to phosphorylation of ADP. Understanding ATP synthase unites bioenergetics, membrane biology, and enzyme mechanics, explaining how cells convert electrochemical energy to biochemical energy, essential for sustaining life processes across all organisms from prokaryotes to complex multicellular life forms.

Q399. Answer: C

Explanation: The cytoskeleton maintains cell shape, supports organelle movement, and enables motility through microfilaments, microtubules, and intermediate filaments. It does not directly synthesize proteins; ribosomes perform that function. Understanding cytoskeletal organization reveals mechanical strength, intracellular transport via motor proteins, and cell signaling. It integrates structural and functional perspectives essential to cellular dynamics, development, and movement in eukaryotic systems across biological research.

Q400. Answer: B

Explanation: The Calvin cycle, or light-independent reaction, occurs in the chloroplast stroma, using ATP and NADPH from light reactions to fix CO_2 into glucose. It regenerates ribulose bisphosphate for continual carbon assimilation. Understanding this biochemical cycle links energy conversion, carbon metabolism, and ecosystem productivity, illustrating photosynthesis' role in sustaining life by forming organic molecules essential to all trophic levels globally.

Questions (401-500)

Q401. The semi-permeable nature of cell membranes allows:

A. Free movement of all molecules
B. Selective transport of specific substances
C. No molecular movement
D. Only water passage

Q402. Which cellular structure contains hydrolytic enzymes for intracellular digestion?

A. Lysosome
B. Golgi apparatus
C. Peroxisome
D. Mitochondrion

Q403. The process by which plant cells lose water through evaporation from leaves is called:

A. Photosynthesis
B. Diffusion
C. Osmosis
D. Transpiration

Q404. What structure provides mechanical strength to plant cells?

A. Cell wall
B. Plasma membrane
C. Nucleus
D. Vacuole

Q405. The main purpose of cellular respiration is to:

A. Convert light energy into glucose
B. Break down glucose to release ATP
C. Store energy as starch
D. Absorb carbon dioxide

Q406. Which organelle detoxifies hydrogen peroxide in cells?

A. Lysosome
B. Peroxisome
C. Golgi apparatus
D. Ribosome

Q407. What structure connects plant cells and allows molecular communication?

A. Tight junctions
B. Plasmodesmata
C. Desmosomes
D. Gap junctions

Q408. The cytoplasmic organelles responsible for photosynthesis are:

A. Ribosomes
B. Lysosomes
C. Chloroplasts
D. Mitochondria

Q409. The organelle responsible for packaging proteins into vesicles for secretion is the:

A. Endoplasmic reticulum
B. Golgi apparatus
C. Nucleolus
D. Ribosome

Q410. What process ensures genetic consistency during cell division?

A. Crossing over
B. Mitosis
C. Meiosis
D. Transcription

Q411. In the cell membrane, cholesterol's main function is to:
A. Provide energy
B. Stabilize membrane fluidity
C. Produce proteins
D. Control ion channels

Q412. The primary structural molecule of plant cell walls is:
A. Cellulose
B. Chitin
C. Collagen
D. Keratin

Q413. The phase of the cell cycle where the cell grows and performs normal functions is:
A. S phase
B. M phase
C. G1 phase
D. G2 phase

Q414. The process of programmed cell death is known as:
A. Autophagy
B. Necrosis
C. Apoptosis
D. Differentiation

Q415. The double membrane enclosing the nucleus separates:
A. Cytoplasm and nucleolus
B. Cytoplasm and nucleoplasm
C. Endoplasmic reticulum and cytosol
D. Nucleolus and Golgi apparatus

Q416. The site of most ATP production during cellular respiration is the:
A. Cytoplasm
B. Ribosome
C. Nucleus
D. Mitochondrial inner membrane

Q417. The term "semi-autonomous organelle" applies to mitochondria and chloroplasts because they:
A. Lack ribosomes
B. Contain their own DNA and ribosomes
C. Are found only in prokaryotes
D. Cannot divide

Q418. The function of the rough endoplasmic reticulum is to:
A. Detoxify chemicals
B. Synthesize membrane-bound proteins
C. Produce lipids
D. Store calcium ions

Q419. In photosynthesis, oxygen is produced from:
A. Carbon dioxide
B. Glucose
C. Water

D. ATP

Q420. The process of endocytosis allows cells to:

A. Expel waste

B. Import large particles or fluids

C. Transport small ions

D. Release hormones

Q421. The central vacuole in plant cells functions to:

A. Produce ATP

B. Synthesize DNA

C. Perform protein synthesis

D. Store water and maintain turgor pressure

Q422. What happens during cytokinesis in animal cells?

A. Nuclear envelope reforms

B. Cytoplasm divides forming two cells

C. Chromosomes condense

D. DNA replicates

Q423. The cytosol primarily consists of:

A. Lipids

B. Proteins and water

C. DNA

D. Starch

Q424. The function of messenger RNA (mRNA) is to:

A. Synthesize proteins directly

B. Carry genetic information from DNA to ribosomes

C. Form ribosomal structure

D. Transport amino acids

Q425. The organelle responsible for converting light energy into chemical energy is the:

A. Mitochondrion

B. Nucleus

C. Chloroplast

D. Golgi apparatus

Q426. The basic unit of heredity that occupies a specific position on a chromosome is called a:

A. Protein

B. Gene

C. Ribosome

D. Codon

Q427. The complementary DNA sequence to 5'-ATGCGT-3' is:

A. 5'-TACGCA-3'

B. 3'-TACGCA-5'

C. 5'-ATGCGT-3'

D. 3'-AUGCGU-5'

Q428. The law of segregation states that:

A. Alleles are inherited together

B. Two alleles for a trait separate during gamete formation

C. Dominant alleles mask recessive ones permanently

D. Traits blend in offspring

Q429. Which process produces gametes with half the number of chromosomes?

A. Meiosis

B. Mitosis
C. Replication
D. Transcription

Q430. A cross between heterozygous tall (Tt) plants produces a genotypic ratio of:

A. 1:1
B. 1:2:1
C. 3:1
D. 2:1:1

Q431. Which molecule carries amino acids to the ribosome during protein synthesis?

A. mRNA
B. tRNA
C. rRNA
D. DNA

Q432. A mutation that changes a single base pair in DNA is called a:

A. Frameshift mutation
B. Point mutation
C. Deletion
D. Inversion

Q433. When both alleles in a heterozygote are fully expressed, the result is:

A. Codominance
B. Incomplete dominance
C. Recessiveness
D. Epistasis

Q434. The observable characteristics of an organism are referred to as its:

A. Genotype
B. Phenotype
C. Allele
D. Chromosome

Q435. Which enzyme unwinds the DNA double helix during replication?

A. Ligase
B. Polymerase
C. Helicase
D. Primase

Q436. A cross between a homozygous dominant and a homozygous recessive organism produces:

A. All dominant phenotype offspring
B. All recessive phenotype offspring
C. A 3:1 ratio
D. A 1:2:1 ratio

Q437. The structure of DNA was discovered by:

A. Mendel and Darwin
B. Watson and Crick
C. Avery and Griffith
D. Franklin and Pauling

Q438. Independent assortment occurs because:

A. Chromosomes replicate before division
B. Homologous chromosomes separate randomly during meiosis I
C. Sister chromatids fail to separate
D. Crossing over never occurs

Q439. Which type of RNA forms the core of the ribosome and catalyzes peptide bond formation?
A. mRNA
B. tRNA
C. rRNA
D. siRNA
Q440. A testcross determines:
A. Chromosome number
B. Genotype of an organism expressing the dominant trait
C. Type of mutation in a gene
D. Sex-linked inheritance
Q441. What is the function of DNA ligase in replication?
A. Synthesizes RNA primers
B. Seals Okazaki fragments on the lagging strand
C. Unwinds the DNA helix
D. Proofreads new DNA strands
Q442. The physical exchange of genetic material between homologous chromosomes is called:
A. Mutation
B. Recombination suppression
C. Segregation
D. Crossing over
Q443. An individual with two identical alleles for a trait is:
A. Heterozygous
B. Homozygous
C. Codominant
D. Epistatic
Q444. The process by which DNA is copied into RNA is called:
A. Translation
B. Replication
C. Transcription
D. Mutation
Q445. Which type of mutation can result in a completely different amino acid sequence downstream?
A. Silent mutation
B. Missense mutation
C. Nonsense mutation
D. Frameshift mutation
Q446. Natural selection acts directly on an organism's:
A. Genes
B. DNA sequences
C. Phenotype
D. Genotype only
Q447. The modern synthesis of evolution combines:
A. Genetics and embryology
B. Darwinian natural selection with Mendelian genetics
C. Mutation theory and vitalism
D. Paleontology and geophysics
Q448. Which evidence best supports common ancestry among species?
A. Similar ecological roles
B. Analogous structures

C. Homologous structures
D. Different DNA sequences

Q449. Genetic drift is most significant in:
A. Large populations
B. Stable environments
C. Small populations
D. Asexual reproduction

Q450. The mechanism by which new species arise is called:
A. Adaptation
B. Speciation
C. Selection
D. Mutation

Q451. Which term describes a change in the genetic composition of a population over generations?
A. Mutation
B. Evolution
C. Variation
D. Adaptation

Q452. The structure of RNA differs from DNA because RNA:
A. Contains thymine instead of uracil
B. Is double-stranded
C. Contains uracil instead of thymine
D. Lacks phosphate groups

Q453. Which of the following represents a sex-linked trait in humans?
A. Skin color
B. Blood type
C. Color blindness
D. Hair texture

Q454. Which scientist proposed the theory of natural selection as a mechanism for evolution?
A. Gregor Mendel
B. Charles Darwin
C. Jean-Baptiste Lamarck
D. Alfred Wallace

Q455. A population in Hardy-Weinberg equilibrium experiences:
A. Frequent mutations
B. No evolutionary change
C. Natural selection
D. Migration

Q456. Which type of natural selection favors individuals with intermediate traits?
A. Stabilizing selection
B. Directional selection
C. Disruptive selection
D. Balancing selection

Q457. The process by which organisms better adapted to their environment tend to survive and reproduce is:
A. Mutation
B. Adaptation
C. Natural selection
D. Genetic drift

Q458. A genetic map shows:
A. Chromosome number
B. Relative positions of genes on chromosomes
C. Protein sequences
D. Mutation rates
Q459. Which process maintains genetic variation within populations?
A. Inbreeding
B. Gene flow
C. Directional selection
D. Stabilizing selection
Q460. When a population's size is drastically reduced by a natural disaster, genetic diversity decreases due to:
A. Mutation
B. Bottleneck effect
C. Gene flow
D. Artificial selection
Q461. Which concept describes small genetic changes accumulating gradually over long periods?
A. Punctuated equilibrium
B. Catastrophism
C. Gradualism
D. Saltation
Q462. Which molecule is responsible for carrying genetic information in most organisms?
A. RNA
B. Protein
C. DNA
D. Lipid
Q463. A diagram that shows evolutionary relationships among organisms is a:
A. Karyotype
B. Phylogenetic tree
C. Gene map
D. Population chart
Q464. A heterozygote advantage maintains polymorphism when:
A. Homozygotes have higher fitness
B. Both alleles are deleterious
C. Heterozygotes possess greater fitness than either homozygote
D. Allele frequencies remain fixed
Q465. Which process results in the formation of gametes with new combinations of alleles?
A. Mitosis
B. Crossing over
C. Transcription
D. Translation
Q466. The term used for genetic differences among individuals within a population is:
A. Variation
B. Mutation
C. Selection
D. Speciation
Q467. Evolutionary fitness measures an organism's:
A. Strength and speed

B. Ability to reproduce successfully
C. Longevity
D. Genetic mutation rate

Q468. Mutations in germ cells are significant because they:
A. Affect only somatic tissues
B. Cannot be inherited
C. Can be passed to offspring
D. Always cause lethal defects

Q469. The fossil record supports evolution by showing:
A. Constant species with no change
B. Sudden appearance without transition
C. Gradual changes and transitional forms
D. Random assortment of organisms

Q470. A mutation that provides an advantage in survival is called a:
A. Neutral mutation
B. Beneficial mutation
C. Lethal mutation
D. Silent mutation

Q471. Which scientist independently proposed a theory of evolution similar to Darwin's?
A. Gregor Mendel
B. Alfred Russel Wallace
C. Thomas Hunt Morgan
D. Charles Lyell

Q472. The process by which organisms migrate and exchange genes between populations is called:
A. Genetic drift
B. Mutation
C. Gene flow
D. Adaptation

Q473. Which factor can cause a population's allele frequencies to change randomly from one generation to the next?
A. Genetic drift
B. Natural selection
C. Gene flow
D. Adaptation

Q474. Evidence for evolution from molecular biology includes:
A. Fossilized remains
B. Homologous protein and DNA sequences
C. Geological strata
D. Vestigial structures only

Q475. Which of the following best illustrates the unity of life across diverse organisms?
A. The presence of different DNA structures in each each species
B. The variation of organelles among eukaryotes
C. The absence of homeostasis in prokaryotes
D. The universality of the genetic code

Q476. Which kingdom contains organisms that are all multicellular and photosynthetic?
A. Animalia
B. Fungi
C. Plantae

D. Protista

Q477. Which structure is common to all cells?

A. Nucleus

B. Cell wall

C. Plasma membrane

D. Chloroplast

Q478. Which statement best explains how classification reflects evolutionary relationships?

A. Organisms are grouped solely by physical similarity.

B. Taxonomy now emphasizes shared ancestry determined by molecular data.

C. Classification ignores genetic information.

D. Evolutionary history does not influence taxonomy.

Q479. What is the primary role of natural selection in biodiversity?

A. Eliminating variation from populations

B. Generating random mutations

C. Favoring traits that enhance survival and reproduction

D. Preventing adaptation

Q480. Which domain includes unicellular organisms that thrive in extreme environments?

A. Bacteria

B. Archaea

C. Eukarya

D. Protista

Q481. Which of the following is true of prokaryotic cells?

A. They contain membrane-bound organelles.

B. Their DNA is enclosed within a nucleus.

C. They lack membrane-bound organelles and have circular DNA.

D. They are exclusively multicellular.

Q482. Which structure in plant cells maintains turgor pressure?

A. Nucleus

B. Cell wall

C. Vacuole

D. Mitochondrion

Q483. Which of the following processes is essential for maintaining genetic diversity in sexually reproducing populations?

A. Binary fission

B. Crossing over during meiosis

C. Mitosis

D. Budding

Q484. Which of the following best describes adaptive radiation?

A. Gradual extinction of species

B. Rapid diversification from a common ancestor

C. Random mutation without environmental influence

D. Loss of genetic variation

Q485. What is the main difference between autotrophs and heterotrophs?

A. Autotrophs consume other organisms; heterotrophs produce their own food.

B. Autotrophs produce food using energy from sunlight or chemicals, while heterotrophs consume organic matter.

C. Both depend directly on sunlight for energy.

D. Both obtain nutrients from decomposition.

Q486. What is a major characteristic distinguishing fungi from plants?
A. Fungi perform photosynthesis; plants absorb nutrients.
B. Fungi absorb nutrients externally through enzymes.
C. Plants reproduce only asexually.
D. Fungi lack cell walls.
Q487. Which of the following best defines homeostasis?
A. Maintenance of a stable internal environment
B. Ability to reproduce
C. Rapid evolutionary change
D. Ability to move
Q488. Which process directly results in speciation?
A. Gene flow
B. Genetic drift
C. Reproductive isolation
D. Migration
Q489. Which kingdom includes organisms that are eukaryotic and absorb nutrients from organic material?
A. Protista
B. Plantae
C. Animalia
D. Fungi
Q490. Which structure allows plant cells to communicate directly with one another?
A. Gap junctions
B. Plasmodesmata
C. Desmosomes
D. Tight junctions
Q491. Which organelle is primarily responsible for energy production in eukaryotic cells?
A. Nucleus
B. Chloroplast
C. Mitochondrion
D. Ribosome
Q492. Which of the following describes a major function of meiosis?
A. Producing identical daughter cells
B. Reducing chromosome number by half
C. Repairing damaged DNA
D. Producing somatic cells
Q493. Which of the following is an example of coevolution?
A. Antibiotic resistance in bacteria
B. Flower and pollinator mutual adaptations
C. Genetic drift in small populations
D. Convergent evolution of wings in bats and insects
Q494. Which feature distinguishes animals from other eukaryotes?
A. Cell walls made of cellulose
B. Absence of membrane-bound organelles
C. Multicellularity with specialized tissues
D. Ability to photosynthesize
Q495. Which of the following best explains convergent evolution?
A. Shared ancestry resulting in similar traits

B. Independent evolution of similar traits in unrelated species
C. Genetic drift eliminating diversity
D. Adaptation through artificial selection

Q496. Which component of an ecosystem recycles nutrients from dead organisms?
A. Producers
B. Consumers
C. Decomposers
D. Herbivores

Q497. What type of symmetry is characteristic of most animals with a distinct head region?
A. Radial symmetry
B. Bilateral symmetry
C. Asymmetry
D. Spherical symmetry

Q498. Which group of organisms are autotrophic, eukaryotic, and mostly multicellular?
A. Protists
B. Fungi
C. Plants
D. Animals

Q499. Which structure in animal cells is responsible for packaging and modifying proteins?
A. Golgi apparatus
B. Lysosome
C. Rough endoplasmic reticulum
D. Ribosome

Q500. What is the primary purpose of phylogenetic trees in biology?
A. To depict feeding relationships
B. To illustrate evolutionary relationships among species
C. To show genetic mutations
D. To classify organisms by physical appearance only

Answer keys and Explanations (401-500)

Q401. Answer: B

Explanation: Cell membranes exhibit selective permeability, controlling which substances enter or leave. Transport proteins regulate ions, nutrients, and waste exchange, maintaining homeostasis. Understanding selective permeability clarifies diffusion, active transport, and signal transduction mechanisms fundamental to physiology and molecular biology. It ensures internal stability and communication, reflecting how structure and composition support cell survival and adaptability in changing environments.

Q402. Answer: A

Explanation: Lysosomes contain hydrolytic enzymes that digest macromolecules, recycle organelles, and defend against pathogens. Their acidic environment optimizes enzymatic breakdown. Understanding lysosomal function reveals cellular renewal and degradation processes, highlighting its role in apoptosis, immunity, and metabolism. It connects biochemistry, pathology, and molecular regulation essential for maintaining cellular integrity and efficient resource recycling in eukaryotic systems.

Q403. Answer: D

Explanation: Transpiration involves water evaporation through stomata, creating negative pressure that pulls water upward from roots via xylem. It regulates temperature and nutrient transport. Understanding transpiration connects plant physiology, hydrology, and gas exchange processes, emphasizing the balance between water conservation and photosynthesis. It demonstrates how plants adapt to environmental conditions to sustain life and maintain global water cycles essential for ecosystems.

Q404. Answer: A

Explanation: The cell wall, composed mainly of cellulose, gives rigidity, protection, and structural support to plant cells. It resists osmotic pressure and defines shape. Understanding cell wall composition links plant biochemistry, growth, and defense mechanisms. It integrates structural biology with ecology, explaining how plant cells maintain integrity, resist stress, and interact mechanically with their environment during development and adaptation.

Q405. Answer: B

Explanation: Cellular respiration breaks down glucose into CO_2 and H_2O, releasing ATP that powers cellular functions. Occurring in mitochondria, it includes glycolysis, Krebs cycle, and electron transport. Understanding respiration connects bioenergetics and metabolism, showing how organisms extract usable energy from food. It emphasizes ATP's role as an energy currency critical for maintaining biological activity, growth, and homeostasis across all living systems.

Q406. Answer: B

Explanation: Peroxisomes contain oxidative enzymes that neutralize hydrogen peroxide (H_2O_2) by converting it into water and oxygen via catalase. They metabolize fatty acids and detoxify harmful substances. Understanding peroxisomal function bridges cell metabolism, oxidative stress management, and lipid regulation. It highlights cellular protection mechanisms that sustain metabolic balance, energy production, and survival under oxidative conditions across eukaryotic organisms in various environments.

Q407. Answer: B

Explanation: Plasmodesmata are microscopic channels that traverse cell walls, enabling direct cytoplasmic exchange between plant cells. They facilitate communication, transport of signaling molecules, and coordination of physiological processes. Understanding plasmodesmata reveals intercellular connectivity and integration of plant tissues. It highlights how plants maintain unity and coordination, supporting development, defense, and metabolic regulation across multicellular systems in plant biology.

Q408. Answer: C

Explanation: Chloroplasts perform photosynthesis, capturing solar energy to convert CO_2 and H_2O into glucose and O_2. They contain chlorophyll pigments and thylakoid membranes driving light reactions. Understanding chloroplasts integrates energy transformation, ecology, and evolution, explaining how plants and algae sustain life's energy flow. Their dual-membrane structure and endosymbiotic origin connect cellular evolution to global carbon and oxygen cycles.

Q409. Answer: B

Explanation: The Golgi apparatus modifies and packages proteins into vesicles for secretion or intracellular use. Through enzymatic modification and sorting, it ensures proteins reach correct destinations. Understanding Golgi function connects molecular transport, secretion pathways, and communication systems in cells. It emphasizes post-translational modification's importance for protein stability, signaling, and coordination of cellular logistics essential to eukaryotic cell physiology.

Q410. Answer: B

Explanation: Mitosis divides a parent cell into two genetically identical daughter cells, preserving chromosome number and genetic integrity. It ensures accurate DNA replication and segregation. Understanding mitosis integrates genetics, development, and cancer biology, emphasizing controlled cell proliferation. It demonstrates how errors can lead to mutation or disease, underlining mitosis as a fundamental mechanism of growth and tissue maintenance across eukaryotic life.

Q411. Answer: B

Explanation: Cholesterol modulates membrane fluidity and permeability by preventing phospholipid packing at low temperatures and restricting movement at high temperatures. It stabilizes the lipid bilayer, ensuring structural integrity. Understanding cholesterol's role connects membrane chemistry, fluid dynamics, and cellular adaptation. It explains how balance in lipid composition supports membrane function, communication, and cellular responses under variable environmental and physiological conditions.

Q412. Answer: A

Explanation: Cellulose, a polysaccharide of β-glucose units, provides tensile strength and rigidity to plant cell walls. It forms microfibrils that resist tension and shape the plant. Understanding cellulose structure highlights carbohydrate biochemistry, plant mechanics, and renewable material applications. It connects molecular organization with ecosystem function, emphasizing cellulose's role in supporting growth, structure, and sustainability across terrestrial biological systems globally.

Q413. Answer: C

Explanation: The G1 phase is part of interphase where cells grow, produce organelles, and prepare for DNA synthesis. It's crucial for metabolic activity and checkpoint regulation. Understanding G1 phase dynamics clarifies growth control, cancer research, and differentiation. It integrates physiology and molecular signaling pathways that determine cell fate, maintaining proper division cycles necessary for organismal development and repair processes.

Q414. Answer: C

Explanation: Apoptosis is an orderly, genetically regulated process eliminating damaged or unnecessary cells without inflammation. It involves caspase activation and DNA fragmentation. Understanding apoptosis connects genetics, immunology, and development, explaining tissue remodeling and defense. It distinguishes controlled cell death from necrosis, emphasizing cellular self-regulation vital for health, disease prevention, and organismal homeostasis across multicellular biological systems.

Q415. Answer: B

Explanation: The nuclear envelope separates the cytoplasm from nucleoplasm, maintaining compartmentalization of transcription and translation. It contains nuclear pores regulating molecular exchange. Understanding this boundary explains genetic protection, RNA transport, and spatial organization within eukaryotic cells. It links molecular architecture to gene regulation and evolutionary complexity, showing how nuclear separation underpins advanced cellular coordination and information management.

Q416. Answer: D

Explanation: The mitochondrial inner membrane hosts the electron transport chain, producing ATP via oxidative phosphorylation. Proton gradients drive ATP synthase, coupling electron flow to energy synthesis. Understanding this process unites biochemistry, physics, and energy transformation principles central to metabolism. It emphasizes how efficient ATP production sustains vital processes, illustrating mitochondria's pivotal role as cellular powerhouses across eukaryotic life.

Q417. Answer: B

Explanation: Mitochondria and chloroplasts possess their own circular DNA and ribosomes, enabling partial genetic independence and self-replication. This supports the endosymbiotic theory of their bacterial origin. Understanding semi-autonomy highlights evolutionary integration, bioenergetics, and organelle inheritance. It connects molecular genetics with cellular evolution, emphasizing collaboration between nuclear and organellar genomes sustaining energy conversion and metabolic specialization in eukaryotic organisms.

Q418. Answer: B

Explanation: The rough ER, studded with ribosomes, synthesizes proteins destined for membranes, secretion, or organelles. It ensures correct folding and transport to the Golgi apparatus. Understanding rough ER function links translation, trafficking, and cellular organization. It demonstrates the coordination of protein biosynthesis and membrane integration vital for maintaining cell architecture and communication in eukaryotic systems.

Q419. Answer: C

Explanation: Oxygen evolves during the light-dependent reactions of photosynthesis when water molecules split in photolysis, releasing electrons, protons, and oxygen gas. This reaction occurs in chloroplast thylakoids. Understanding oxygen production connects redox chemistry, energy flow, and ecological cycles sustaining life. It highlights how photosynthesis drives atmospheric balance, global energy transfer, and ecosystem productivity fundamental to planetary health.

Q420. Answer: B

Explanation: Endocytosis enables cells to internalize large molecules or fluids by engulfing them into vesicles formed from the plasma membrane. Types include phagocytosis and pinocytosis. Understanding endocytosis clarifies nutrient uptake, immune defense, and receptor-mediated signaling. It bridges membrane dynamics, cytoskeletal mechanics, and cellular regulation, demonstrating how cells interact with their environment to sustain growth and metabolic exchange.

Q421. Answer: D

Explanation: The central vacuole stores water, nutrients, and waste while maintaining turgor pressure that supports plant rigidity and growth. It aids in detoxification and storage. Understanding vacuole function connects plant anatomy, osmoregulation, and growth mechanics. It emphasizes cellular adaptation to hydration changes, providing insight into how plants regulate internal environment and sustain structure under varying environmental conditions.

Q422. Answer: B

Explanation: Cytokinesis divides the cytoplasm following mitosis, forming two separate daughter cells. In animal cells, a contractile ring of actin and myosin creates a cleavage furrow that pinches the cell membrane. Understanding cytokinesis clarifies cell cycle completion and structural dynamics. It links cytoskeletal function to tissue growth, ensuring proper cellular reproduction and organization within multicellular organisms.

Q423. Answer: B

Explanation: The cytosol is a gel-like matrix composed mainly of water, proteins, ions, and metabolites supporting biochemical reactions. It provides the medium for enzyme activity and organelle suspension. Understanding cytosolic composition reveals metabolic coordination and molecular mobility, illustrating how intracellular environments facilitate regulation and energy flow essential for maintaining cellular homeostasis and biochemical reactions.

Q424. Answer: B

Explanation: Messenger RNA transmits genetic instructions from DNA in the nucleus to ribosomes for protein synthesis. It serves as an intermediary in gene expression, translating genetic code into functional polypeptides. Understanding mRNA function connects molecular genetics, transcription, and translation, explaining how information flow sustains life processes, guiding biotechnology and medical advances such as mRNA vaccines and genetic therapies.

Q425. Answer: C

Explanation: Chloroplasts capture sunlight through chlorophyll and convert it into chemical energy stored as glucose via photosynthesis. They contain thylakoid membranes and stroma for light and dark reactions. Understanding chloroplast function unites biochemistry, ecology, and evolution, demonstrating energy transformation vital for life. It shows how autotrophs sustain biospheric energy flow and carbon cycling supporting ecosystems worldwide.

Q426. Answer: B

Explanation: A gene is a segment of DNA that encodes information for the synthesis of proteins or functional RNA molecules. It determines heritable traits by specifying amino acid sequences through genetic code. Understanding gene structure and function links DNA replication, transcription, and translation, forming the basis for inheritance, gene regulation, and evolution across species through molecular and classical genetics principles.

Q427. Answer: B

Explanation: DNA strands are antiparallel and complementary, pairing adenine with thymine and cytosine with guanine. Thus, the complement of 5'-ATGCGT-3' is 3'-TACGCA-5'. Understanding base pairing ensures accurate replication and transcription, maintaining genetic fidelity. This molecular precision supports inheritance and genetic engineering, illustrating the elegance of the Watson-Crick model in maintaining life's continuity across generations.

Q428. Answer: B

Explanation: Mendel's law of segregation explains that allele pairs separate during meiosis so each gamete receives one allele per gene. This ensures genetic variation and predictable inheritance ratios. Understanding this law underpins Punnett squares, genotype predictions, and genetic counseling. It forms the cornerstone of classical genetics, linking cellular processes with observable inheritance patterns in all sexually reproducing organisms.

Q429. Answer: A

Explanation: Meiosis reduces chromosome number by half, producing haploid gametes through two divisions following one DNA replication. It introduces genetic diversity via crossing over and independent assortment. Understanding meiosis bridges cytogenetics and evolutionary biology, explaining how sexual reproduction maintains chromosome stability across generations while promoting genetic variation critical for adaptation and evolution.

Q430. Answer: B

Explanation: Crossing two heterozygous (Tt) plants yields genotypes TT, Tt, and tt in a 1:2:1 ratio, representing homozygous dominant, heterozygous, and homozygous recessive offspring. Phenotypically, tallness (dominant) appears in 3:1 ratio. Understanding these ratios connects probability with biological inheritance, reinforcing Mendel's foundational work in predicting genetic outcomes essential for studying heredity and population variation.

Q431. Answer: B

Explanation: Transfer RNA (tRNA) transports specific amino acids to ribosomes, matching anticodons with mRNA codons during translation. This ensures accurate protein synthesis according to genetic instructions. Understanding tRNA function bridges molecular genetics and biochemistry, demonstrating how nucleic acid sequences are converted into functional proteins vital for structure, enzymes, and regulation in all living organisms.

Q432. Answer: B

Explanation: A point mutation alters a single nucleotide base pair, potentially changing an amino acid or being silent. Its effects range from neutral to harmful, depending on the mutation's location. Understanding point mutations explains genetic diseases, evolution, and molecular diagnostics. It links DNA replication errors and mutagen exposure to genetic diversity and medical research in genomics.

Q433. Answer: A

Explanation: Codominance occurs when both alleles contribute equally to phenotype, such as AB blood type. Unlike incomplete dominance, where blending occurs, codominant traits show both characteristics distinctly. Understanding codominance reveals genetic interactions beyond simple dominance, enriching knowledge of molecular expression patterns and inheritance diversity in population genetics and evolutionary adaptation mechanisms.

Q434. Answer: B

Explanation: Phenotype represents the physical and physiological traits expressed by an organism's genotype interacting with the environment. Examples include height, eye color, and enzyme activity. Understanding phenotype connects genetic coding with biochemical processes, highlighting the relationship between genes and environment shaping organismal diversity fundamental to evolution and applied genetics fields.

Q435. Answer: C

Explanation: DNA helicase unwinds the double helix at replication forks, separating strands to allow polymerase access. Its activity requires ATP hydrolysis. Understanding helicase function clarifies replication dynamics, repair mechanisms, and genetic stability. It integrates molecular enzymology with replication fidelity, ensuring accurate inheritance of genetic information across cellular divisions and generations in all living systems.

Q436. Answer: A

Explanation: Crossing homozygous dominant (AA) and homozygous recessive (aa) individuals yields heterozygous offspring (Aa), all expressing the dominant phenotype. Understanding monohybrid crosses illustrates dominant-recessive relationships and inheritance predictability. This foundational concept in classical genetics underpins breeding programs, trait analysis, and population genetics used across biology, agriculture, and evolutionary research fields for genetic pattern understanding.

Q437. Answer: B

Explanation: James Watson and Francis Crick proposed the double-helix model of DNA in 1953, using X-ray diffraction data from Rosalind Franklin. Their discovery revealed complementary base pairing and antiparallel strands. Understanding DNA structure revolutionized molecular biology, linking heredity, replication, and gene expression to macromolecular architecture essential for modern genetics, biotechnology, and evolutionary research worldwide.

Q438. Answer: B

Explanation: During meiosis I, homologous chromosomes align independently, producing gametes with random allele combinations. This principle of independent assortment increases genetic diversity among offspring. Understanding this mechanism integrates cytogenetics and population genetics, highlighting how meiotic behavior drives variation, evolution, and inheritance patterns essential for species adaptability and natural selection processes in biological systems.

Q439. Answer: C

Explanation: Ribosomal RNA (rRNA) constitutes the structural and catalytic core of ribosomes, facilitating peptide bond formation during translation. It interacts with tRNA and mRNA to ensure precise protein synthesis. Understanding rRNA's enzymatic role (ribozyme) reveals molecular evolution insights, showing how RNA predates proteins in catalysis and highlighting ribosome's central role in gene expression and cellular biochemistry.

Q440. Answer: B

Explanation: A testcross involves crossing an individual with a dominant phenotype and an unknown genotype with a homozygous recessive partner. The offspring ratios reveal the dominant individual's genotype. Understanding testcrossing clarifies Mendelian inheritance, probability, and genotype-phenotype relationships. It remains vital for genetics education, breeding strategies, and identifying heterozygosity in research and applied genetics fields.

Q441. Answer: B

Explanation: DNA ligase joins Okazaki fragments on the lagging strand by forming phosphodiester bonds, sealing nicks between fragments. Its action ensures a continuous DNA strand post-replication. Understanding ligase activity clarifies replication mechanics and molecular repair processes. It's essential for maintaining genome stability and underpins biotechnology applications like recombinant DNA cloning and genetic engineering advancements globally.

Q442. Answer: D

Explanation: Crossing over occurs during prophase I of meiosis, when homologous chromosomes exchange DNA segments, producing recombinant chromosomes. This increases genetic variation and evolutionary potential. Understanding crossing over connects molecular recombination, inheritance mapping, and diversity. It highlights meiosis's role in evolution, showing how chromosomal behavior generates novel gene combinations essential for adaptation and natural selection.

Q443. Answer: B

Explanation: Homozygous individuals possess identical alleles for a gene, either dominant (AA) or recessive (aa). This genetic uniformity results in predictable phenotypes. Understanding homozygosity and heterozygosity provides insight into inheritance patterns, population genetics, and selective breeding. It connects genotype-phenotype relationships and underlies genetic stability or diversity influencing adaptation and evolutionary processes across biological populations.

Q444. Answer: C

Explanation: Transcription converts DNA-encoded genetic information into messenger RNA using RNA polymerase. This process occurs in the nucleus of eukaryotes and the cytoplasm of prokaryotes. Understanding transcription bridges genetics and biochemistry, explaining how gene expression begins and how RNA mediates protein synthesis. It forms a foundation for understanding regulation, gene expression control, and molecular biology techniques.

Q445. Answer: D

Explanation: Frameshift mutations occur from nucleotide insertions or deletions not in multiples of three, altering the reading frame. This often leads to nonfunctional proteins. Understanding frameshift effects clarifies mutagenesis, disease causation, and molecular evolution. It connects DNA sequence integrity with protein functionality and underscores the importance of replication accuracy in maintaining organismal health and genetic stability.

Q446. Answer: C

Explanation: Natural selection operates on phenotypic traits that affect survival and reproduction, indirectly influencing genotypic frequencies over time. Favorable phenotypes become more common in populations. Understanding this principle integrates genetics, ecology, and evolution, showing how environmental pressures shape species diversity. It emphasizes adaptation mechanisms driving biodiversity and long-term evolutionary change across ecosystems and populations globally.

Q447. Answer: B

Explanation: The modern synthesis unites Darwin's natural selection with Mendelian inheritance, explaining evolution as changes in allele frequencies within populations. It incorporates mutation, genetic drift, and gene flow. Understanding this framework integrates molecular biology, paleontology, and population genetics, providing a unified explanation of evolutionary mechanisms and biodiversity patterns that define modern evolutionary theory across biological disciplines.

Q448. Answer: C

Explanation: Homologous structures share anatomical and genetic origins despite functional differences, indicating descent from common ancestors. Examples include vertebrate limbs. Understanding homology combines comparative anatomy, embryology, and molecular biology, revealing evolutionary relationships and diversification patterns. It supports phylogenetic analysis and evolutionary inference central to understanding species origin and classification across the tree of life.

Q449. Answer: C

Explanation: Genetic drift causes random changes in allele frequencies, strongly impacting small populations where chance events can fix or eliminate alleles. It reduces genetic variation and may counteract selection. Understanding drift elucidates founder and bottleneck effects, highlighting stochastic influences on evolution. It integrates population genetics and conservation biology, demonstrating how randomness shapes diversity and evolution over generations.

Q450. Answer: B

Explanation: Speciation occurs when populations diverge genetically and reproductively, forming distinct species. Mechanisms include geographic isolation, genetic drift, and natural selection. Understanding speciation integrates genetics, ecology, and evolutionary biology, illustrating biodiversity's dynamic nature. It emphasizes how reproductive barriers evolve, driving the branching pattern of life central to macroevolution and the origin of new taxa on Earth.

Q451. Answer: B

Explanation: Evolution is the gradual change in a population's genetic composition across generations due to natural selection, mutation, genetic drift, and gene flow. It leads to adaptation and species diversification. Understanding evolution integrates molecular biology, genetics, and ecology, explaining biodiversity and common ancestry. It forms the foundation of biological sciences and supports interpretation of fossil, genetic, and anatomical evidence across species.

Q452. Answer: C

Explanation: RNA contains uracil instead of thymine and is usually single-stranded, allowing it to perform various roles such as catalysis, coding, and regulation. Understanding RNA's chemical composition clarifies gene expression and translation. It reveals RNA's versatility in evolution and cellular control, linking molecular structure to function, gene regulation, and modern genetic technologies like RNA sequencing and therapeutic development.

Q453. Answer: C

Explanation: Color blindness is a sex-linked trait carried on the X chromosome, often recessive, explaining its higher prevalence in males. Understanding sex linkage clarifies inheritance differences between sexes and reveals chromosome-specific gene expression. It integrates cytogenetics and medical genetics, providing insight into hereditary disorders and X-linked inheritance mechanisms important in diagnosing and studying genetic diseases.

Q454. Answer: B

Explanation: Charles Darwin proposed natural selection in 1859, describing how heritable variations that improve survival or reproduction become more common. His observations of finches and diverse species formed evolutionary theory foundations. Understanding Darwin's concept connects adaptation, genetics, and ecology, explaining evolution through selection pressure and variation central to biodiversity and evolutionary biology worldwide.

Q455. Answer: B

Explanation: In Hardy-Weinberg equilibrium, allele and genotype frequencies remain constant over generations when conditions such as random mating and no selection, mutation, or migration are met. Understanding this principle establishes a baseline for detecting evolutionary forces. It integrates population genetics, mathematics, and molecular analysis to measure evolution's magnitude and mechanisms across biological populations over time.

Q456. Answer: A

Explanation: Stabilizing selection favors intermediate phenotypes, reducing variation and maintaining population stability around an average trait. It often occurs in stable environments. Understanding stabilizing selection reveals evolutionary balance between extremes and adaptation efficiency. It integrates ecology, genetics, and physiology, highlighting mechanisms preserving optimal traits and contributing to long-term population resilience across species.

Q457. Answer: C

Explanation: Natural selection is the differential survival and reproduction of individuals based on inherited traits suited to environmental conditions. It drives evolution by increasing beneficial alleles. Understanding natural selection connects genetics, ecology, and molecular evolution, illustrating adaptation and species divergence. It underpins evolutionary theory, conservation, and applied biology, explaining biodiversity and organismal fitness across environments.

Q458. Answer: B

Explanation: Genetic maps indicate gene locations based on recombination frequencies during meiosis, measured in map units or centimorgans. Understanding mapping reveals gene linkage, inheritance patterns, and chromosomal organization. It supports genomic research, medical diagnostics, and evolutionary studies by correlating genetic distance with recombination probability, linking molecular and classical genetics in mapping traits and hereditary variation.

Q459. Answer: B

Explanation: Gene flow, or migration, introduces alleles from one population to another, maintaining or increasing genetic diversity. It counteracts genetic drift and local adaptation. Understanding gene flow integrates ecology, genetics, and evolution, showing how connectivity among populations influences adaptation, speciation, and conservation management in maintaining biodiversity across ecosystems and species over time.

Q460. Answer: B

Explanation: The bottleneck effect occurs when population reduction randomly eliminates alleles, reducing genetic variability. It leads to inbreeding and decreased adaptability. Understanding this effect combines population genetics and conservation biology, highlighting how chance events shape evolution. It emphasizes protecting endangered species' gene pools to preserve evolutionary potential and biodiversity in ecological and environmental management contexts.

Q461. Answer: C

Explanation: Gradualism proposes that evolution proceeds through slow, continuous genetic changes leading to major differences over time. It contrasts punctuated equilibrium, which suggests bursts of rapid change. Understanding gradualism clarifies fossil record interpretation, evolutionary rate studies, and long-term adaptation. It connects paleontology, genetics, and evolutionary theory, explaining how incremental changes drive species development across geological timescales.

Q462. Answer: C

Explanation: DNA encodes hereditary information through nucleotide sequences forming genes that determine cellular functions. Its double-helix structure ensures stability and precise replication. Understanding DNA's role connects molecular genetics and evolutionary biology, explaining inheritance, mutation, and variation. It underpins biotechnology, genomics, and medical research, highlighting DNA as life's blueprint and evolutionary record across all species globally.

Q463. Answer: B

Explanation: A phylogenetic tree illustrates evolutionary relationships based on shared ancestry, morphology, and molecular data. Branching patterns reflect divergence events. Understanding phylogenies integrates systematics, molecular biology, and paleontology, providing frameworks for taxonomy and comparative genomics. It helps trace lineage relationships and evolutionary trends, uniting historical biology and modern genetics in biodiversity interpretation and evolutionary analysis.

Q464. Answer: C

Explanation: Heterozygote advantage occurs when heterozygotes exhibit superior fitness, maintaining multiple alleles in a population. Example: sickle-cell trait confers malaria resistance. Understanding this concept links population genetics and evolutionary medicine, explaining how natural selection preserves genetic variation balancing adaptive benefits and costs. It demonstrates equilibrium between alleles influencing health and evolution across species.

Q465. Answer: B

Explanation: Crossing over during meiosis exchanges DNA segments between homologous chromosomes, generating recombinant gametes. It increases genetic variation, fueling evolution and adaptation. Understanding recombination unites molecular and evolutionary genetics, explaining inheritance mapping and diversity origins. It illustrates meiosis's role in producing variability critical to evolution, agriculture, and genetic studies across living organisms.

Q466. Answer: A

Explanation: Genetic variation encompasses differences in alleles and genotypes among individuals, arising through mutation, recombination, and gene flow. It enables populations to adapt to changing environments. Understanding variation connects genetics, ecology, and evolution, highlighting the raw material for selection and drift. It explains diversity within species and supports conservation, agriculture, and medical research advancements.

Q467. Answer: B

Explanation: Evolutionary fitness quantifies reproductive success, reflecting how well an organism passes its genes to the next generation. Survival alone does not ensure fitness without successful reproduction. Understanding fitness links genetics, ecology, and evolutionary biology, emphasizing adaptation effectiveness and selective pressures shaping populations. It underpins evolutionary models and natural selection analysis in population studies.

Q468. Answer: C

Explanation: Germ cell mutations occur in reproductive cells and can be transmitted to progeny, influencing genetic diversity and evolution. Somatic mutations affect only individual organisms. Understanding germline mutations connects molecular biology, heredity, and evolutionary theory, explaining how variation arises across generations and contributes to adaptation, disease, and evolutionary innovation within populations globally.

Q469. Answer: C

Explanation: Fossil evidence demonstrates gradual morphological changes and transitional species linking ancestral and modern forms. It corroborates evolutionary patterns through geological time. Understanding fossil data integrates paleontology, geology, and evolutionary biology, providing direct evidence for descent with modification. It illustrates species adaptation, extinction, and diversification driving Earth's biological history across millions of years.

Q470. Answer: B

Explanation: Beneficial mutations improve an organism's fitness, enhancing survival or reproduction under specific environmental conditions. They drive adaptation and natural selection. Understanding beneficial mutations links molecular genetics and evolution, showing how genetic changes become fixed in populations. It highlights mutation's dual role in generating variation and enabling adaptive evolution critical to species resilience and diversity.

Q471. Answer: B

Explanation: Alfred Russel Wallace formulated a theory of evolution by natural selection independently of Charles Darwin, prompting Darwin to publish his work. Wallace's studies on biogeography supported shared mechanisms of adaptation. Understanding Wallace's contribution highlights collaborative scientific discovery and how comparative studies across ecosystems reveal evolutionary patterns and species distribution globally through natural selection.

Q472. Answer: C

Explanation: Gene flow occurs when individuals or gametes move between populations, transferring alleles and maintaining genetic connectivity. It prevents divergence and can introduce adaptive traits. Understanding gene flow integrates population genetics, ecology, and conservation, emphasizing how migration influences diversity, adaptation, and speciation in evolutionary and environmental contexts across interrelated biological systems.

Q473. Answer: A

Explanation: Genetic drift is random fluctuation of allele frequencies due to chance events, especially in small populations. It may lead to loss of alleles and reduced diversity. Understanding drift connects stochastic processes to evolution, illustrating how randomness complements selection in shaping populations. It underpins population genetics and evolutionary dynamics explaining variability and divergence in species evolution.

Q474. Answer: B

Explanation: Molecular evidence from homologous DNA and protein sequences reveals shared ancestry, with closer relationships showing higher sequence similarity. Comparative genomics confirms evolutionary links beyond morphology. Understanding molecular homology integrates biochemistry and evolution, providing precise phylogenetic insights. It reinforces evolutionary theory, demonstrating that all life shares a common genetic foundation reflected in conserved molecular structures across species.

Q475. Answer: D

Explanation: All known organisms use DNA with the same four nucleotide bases and a nearly universal genetic code, reflecting a shared evolutionary origin. This molecular uniformity underpins the principle that all life on Earth descended from common ancestors, illustrating evolutionary continuity through genetic mechanisms that transcend species boundaries.

Q476. Answer: C

Explanation: Members of the kingdom Plantae are multicellular eukaryotes that perform photosynthesis using chlorophyll within chloroplasts. They form the foundation of terrestrial ecosystems by converting sunlight into chemical energy, supporting food webs and maintaining oxygen levels critical to life.

Q477. Answer: C

Explanation: The plasma membrane, composed of a phospholipid bilayer with embedded proteins, is present in all cells, both prokaryotic and eukaryotic. It controls the passage of substances and maintains homeostasis, demonstrating a fundamental organizational feature of all living organisms.

Q478. Answer: B

Explanation: Modern taxonomy integrates phylogenetic analyses and molecular sequencing to identify common ancestry. DNA comparisons clarify evolutionary pathways, aligning classification with evolutionary descent rather than superficial morphological resemblance, making taxonomy more accurate and reflective of evolutionary history.

Q479. Answer: C

Explanation: Natural selection acts on existing variation, favoring traits that increase fitness in specific environments. Over generations, these advantageous traits become more common, promoting adaptation and contributing to the vast diversity of life seen in different ecological contexts.

Q480. Answer: B

Explanation: Archaea consist of prokaryotic microorganisms adapted to extreme conditions like high salinity, temperature, or acidity. Their unique biochemical pathways and membrane lipids distinguish them from bacteria and highlight evolutionary adaptations to diverse habitats.

Q481. Answer: C

Explanation: Prokaryotic cells, including bacteria and archaea, lack nuclei and internal membranes, with genetic material arranged in circular DNA. Their simplicity and efficiency reflect early evolutionary adaptations that allowed them to colonize diverse environments.

Q482. Answer: C

Explanation: The large central vacuole stores water and solutes, maintaining turgor pressure that supports plant rigidity. This pressure ensures upright growth and plays a vital role in nutrient storage, cellular homeostasis, and structural stability.

Q483. Answer: B

Explanation: During meiosis, homologous chromosomes exchange genetic segments through crossing over, generating novel allele combinations. This genetic recombination enhances variation, fueling evolutionary potential and adaptability in changing environments.

Q484. Answer: B

Explanation: Adaptive radiation occurs when one ancestral species diversifies into multiple forms, each adapted to unique ecological niches. This process drives biodiversity, as seen in Darwin's finches, demonstrating evolutionary responsiveness to environmental opportunities.

Q485. Answer: B

Explanation: Autotrophs synthesize organic compounds from inorganic sources using photosynthesis or chemosynthesis, forming the base of food webs. Heterotrophs rely on consuming organic material, establishing energy flow and interdependence in ecosystems.

Q486. Answer: B

Explanation: Fungi are heterotrophic decomposers that secrete enzymes to digest organic matter externally, absorbing nutrients afterward. Their chitinous cell walls and absorptive nutrition contrast with plants' cellulose walls and photosynthetic metabolism.

Q487. Answer: A

Explanation: Homeostasis refers to dynamic regulatory processes that maintain internal stability despite external fluctuations, crucial for survival. Mechanisms such as temperature regulation and osmoregulation exemplify this principle across animal and plant physiology.

Q488. Answer: C

Explanation: Speciation occurs when populations become reproductively isolated, preventing gene exchange. Over time, genetic differences accumulate, leading to the emergence of distinct species, a fundamental mechanism driving biological diversity and evolution.

Q489. Answer: D

Explanation: Fungi are eukaryotic heterotrophs that absorb nutrients by secreting enzymes to decompose organic matter. This absorption-based nutrition distinguishes them from animals and plants, making them essential decomposers in ecosystems.

Q490. Answer: B

Explanation: Plasmodesmata are cytoplasmic channels through cell walls that enable the transport of molecules and signaling between plant cells. This connectivity integrates cellular functions, facilitating coordinated growth and physiological responses.

Q491. Answer: C

Explanation: Mitochondria generate ATP through cellular respiration, converting biochemical energy from nutrients into usable energy. They possess their own DNA, reflecting endosymbiotic origins and underscoring energy transformation as a universal biological process.

Q492. Answer: B

Explanation: Meiosis produces haploid gametes by halving chromosome number, ensuring genetic stability across generations. This reduction, coupled with recombination, maintains species chromosome counts while promoting genetic diversity crucial for evolution.

Q493. Answer: B

Explanation: Coevolution occurs when two species reciprocally influence each other's evolution, such as flowers evolving traits to attract specific pollinators. This interdependence shapes biodiversity and ecological relationships across ecosystems.

Q494. Answer: C

Explanation: Animals are multicellular eukaryotes exhibiting cellular differentiation and complex tissue specialization. This organization enables locomotion, sensory perception, and complex behaviors, defining the animal kingdom's functional and structural diversity.

Q495. Answer: B

Explanation: Convergent evolution arises when unrelated species evolve similar adaptations due to comparable environmental pressures. Examples like wings in bats and birds demonstrate evolutionary solutions shaped by function rather than ancestry.

Q496. Answer: C

Explanation: Decomposers, including fungi and bacteria, break down dead organic matter, releasing nutrients into the environment. This recycling supports ecosystem sustainability, connecting energy flow and matter cycling within the biosphere.

Q497. Answer: B

Explanation: Bilateral symmetry allows for directional movement and cephalization, facilitating sensory coordination and predation. This body plan, seen in most animal phyla, reflects an evolutionary adaptation to active locomotion and complex environmental interactions.

Q498. Answer: C

Explanation: Plants synthesize their own food through photosynthesis, converting sunlight into energy-storing carbohydrates. Their multicellularity and autotrophy make them primary producers sustaining terrestrial and aquatic food chains globally.

Q499. Answer: A

Explanation: The Golgi apparatus modifies, sorts, and packages proteins and lipids for transport within or outside the cell. Its role in protein processing and secretion underpins cellular communication and homeostasis in multicellular organisms.

Q500. Answer: B

Explanation: Phylogenetic trees visually represent hypotheses about evolutionary relationships, showing common ancestry and divergence patterns. They integrate molecular and morphological data, providing a framework to trace lineage histories and evolutionary processes.

Questions (501-600)

Q501. Which organelle is found in plant cells but not in animal cells and is responsible for photosynthesis?

A. Mitochondrion
B. Ribosome
C. Chloroplast
D. Lysosome

Q502. Which factor most directly drives evolution within a population?

A. Mutation introducing genetic variation
B. Absence of natural selection
C. Environmental uniformity
D. Genetic stagnation

Q503. Which type of reproduction results in genetically identical offspring?

A. Sexual reproduction
B. Asexual reproduction
C. Cross-pollination
D. Fertilization

Q504. Which characteristic is shared by all members of the animal kingdom?

A. Photosynthetic capability
B. Multicellularity and heterotrophy
C. Cell walls made of cellulose
D. Ability to fix nitrogen

Q505. Which process maintains genetic stability across generations in sexually reproducing organisms?

A. Mutation
B. Meiosis and fertilization
C. Mitosis
D. Gene flow

Q506. Which evidence best supports the theory of evolution?

A. Identical fossil sequences worldwide
B. Fossil records showing gradual changes in related species
C. Lack of genetic variation in populations
D. Sudden disappearance of all species

Q507. Which level of biological classification is the most specific?

A. Kingdom
B. Phylum
C. Genus
D. Species

Q508. Which of the following describes commensalism?

A. Both organisms benefit
B. One organism benefits, the other is unaffected

C. One organism is harmed
D. Both organisms are harmed

Q509. Which type of macromolecule primarily serves as enzymes?
A. Proteins
B. Lipids
C. Nucleic acids
D. Carbohydrates

Q510. Which process is directly responsible for oxygen production in photosynthesis?
A. Calvin cycle
B. Light-dependent reactions
C. Glycolysis
D. Electron transport in mitochondria

Q511. Which factor distinguishes vascular from nonvascular plants?
A. Presence of roots only
B. Presence of conducting tissues xylem and phloem
C. Ability to reproduce sexually
D. Possession of chlorophyll

Q512. Which reproductive strategy ensures genetic variation?
A. Binary fission
B. Budding
C. Sexual reproduction
D. Spore formation

Q513. What is the main function of ribosomes?
A. ATP production
B. Protein synthesis
C. Lipid storage
D. Genetic replication

Q514. Which type of natural selection favors individuals with intermediate traits?
A. Stabilizing selection
B. Directional selection
C. Disruptive selection
D. Artificial selection

Q515. What is the primary function of DNA in living organisms?
A. Transporting proteins
B. Storing and transmitting genetic information
C. Catalyzing metabolic reactions
D. Forming cell membranes

Q516. Which evolutionary mechanism results from random changes in allele frequency, especially in small populations?
A. Gene flow
B. Genetic drift
C. Mutation
D. Natural selection

Q517. What adaptation enables desert plants to minimize water loss?
A. Broad leaves with large stomata
B. Waxy cuticle and CAM photosynthesis
C. Absence of roots
D. Increased transpiration

Q518. Which of the following best defines biodiversity?
A. The number of individuals in a population
B. The variety of species, genes, and ecosystems within an area
C. The dominance of one species in an ecosystem
D. The presence of invasive species
Q519. Which organ system in animals is primarily responsible for maintaining fluid balance?
A. Digestive system
B. Circulatory system
C. Excretory system
D. Reproductive system
Q520. Which process increases the number of cells in multicellular organisms?
A. Meiosis
B. Binary fission
C. Mitosis
D. Differentiation
Q521. Which of the following is an example of behavioral adaptation?
A. Development of thick fur
B. Migration during seasonal changes
C. Possession of gills
D. Growth of protective spines
Q522. What is the primary role of decomposers in ecosystems?
A. Fixing nitrogen
B. Recycling nutrients from organic matter
C. Producing primary energy
D. Regulating photosynthesis
Q523. Which evolutionary pattern results when unrelated species develop similar traits due to comparable environments?
A. Divergent evolution
B. Convergent evolution
C. Coevolution
D. Parallel evolution
Q524. Which organelle is responsible for digesting waste and cellular debris in eukaryotic cells?
A. Golgi apparatus
B. Lysosome
C. Nucleus
D. Endoplasmic reticulum
Q525. Which of the following best explains the significance of genetic variation in populations?
A. It decreases adaptability
B. It allows populations to survive environmental changes
C. It leads to extinction
D. It prevents evolution
Q526. Which of the following best defines an ecosystem?
A. The total number of species in a biome
B. A community of organisms interacting with their physical environment
C. A single population in an area
D. The nonliving components of the biosphere
Q527. The main source of energy for nearly all ecosystems is:
A. The moon

B. The sun
C. The Earth's core
D. Chemical reactions in soil
Q528. The term "biotic factors" refers to:
A. Nonliving elements like sunlight and water
B. Living components such as plants, animals, and microbes
C. Abiotic components that affect ecosystems
D. Chemical elements cycling through the atmosphere
Q529. Which process is primarily responsible for recycling nutrients in ecosystems?
A. Photosynthesis
B. Decomposition
C. Respiration
D. Evaporation
Q530. In a food chain, primary consumers are typically:
A. Herbivores
B. Carnivores
C. Decomposers
D. Producers
Q531. Which biogeochemical cycle involves nitrogen-fixing bacteria converting atmospheric nitrogen into usable forms?
A. Carbon cycle
B. Nitrogen cycle
C. Phosphorus cycle
D. Water cycle
Q532. Which ecosystem typically has the greatest biodiversity?
A. Tundra
B. Desert
C. Tropical rainforest
D. Taiga
Q533. The carrying capacity of an ecosystem is defined as:
A. The maximum population size the environment can sustain
B. The total number of species in a biome
C. The reproductive potential of a population
D. The density of predators in a habitat
Q534. Which trophic level receives the least amount of energy in an ecosystem?
A. Producers
B. Primary consumers
C. Secondary consumers
D. Tertiary consumers
Q535. A biome characterized by low precipitation, extreme temperature variation, and sparse vegetation is a:
A. Desert
B. Grassland
C. Tundra
D. Temperate forest
Q536. The major abiotic factors that determine a biome's characteristics include:
A. Temperature and precipitation
B. Food availability and predation

C. Competition and symbiosis
D. Genetic variation and adaptation

Q537. Which process describes water movement through evaporation, condensation, and precipitation?
A. Nitrogen cycle
B. Carbon cycle
C. Water cycle
D. Oxygen cycle

Q538. In ecological succession, the first organisms to colonize a barren environment are called:
A. Pioneer species
B. Climax species
C. Secondary consumers
D. Opportunistic species

Q539. In a stable ecosystem, the total biomass of producers compared to consumers is:
A. Greater
B. Smaller
C. Equal
D. Variable depending on temperature

Q540. A keystone species is important because it:
A. Always occupies the lowest trophic level
B. Has little effect on ecosystem structure
C. Disproportionately influences community stability
D. Is always the top predator

Q541. Which of the following best describes commensalism?
A. Both species benefit
B. One benefits while the other is unaffected
C. One benefits and the other is harmed
D. Both species are harmed

Q542. The main greenhouse gas produced by burning fossil fuels is:
A. Nitrogen
B. Methane
C. Carbon dioxide
D. Ozone

Q543. The flow of energy through an ecosystem can best be represented by a:
A. Food web
B. Pyramid of numbers
C. Biomass chart
D. Energy pyramid

Q544. Which factor most directly limits the carrying capacity of a population?
A. Predator-prey ratios
B. Availability of resources
C. Seasonal migration
D. Temperature variations

Q545. A species that is introduced to a new ecosystem and causes harm is called a:
A. Native species
B. Endemic species
C. Invasive species
D. Indicator species

Q546. Which process converts atmospheric carbon into organic molecules by plants?
A. Respiration
B. Photosynthesis
C. Combustion
D. Decomposition
Q547. Which population growth model represents limited resources and environmental resistance?
A. Exponential growth
B. Logistic growth
C. Geometric growth
D. Linear growth
Q548. Which organisms are responsible for converting atmospheric nitrogen into ammonia in the soil?
A. Fungi
B. Nitrogen-fixing bacteria
C. Protozoa
D. Algae
Q549. The term "r-strategist" refers to organisms that:
A. Produce few offspring with high parental care
B. Produce many offspring with low survival rates
C. Maintain stable population sizes
D. Are top predators
Q550. The region of Earth that supports all life is known as the:
A. Lithosphere
B. Atmosphere
C. Hydrosphere
D. Biosphere
Q551. Which term describes the gradual change in species composition of a community following disturbance?
A. Adaptation
B. Succession
C. Evolution
D. Mutation
Q552. Which biome is dominated by coniferous forests and experiences long, cold winters?
A. Desert
B. Savanna
C. Taiga
D. Temperate grassland
Q553. Which of the following best defines a food web?
A. Linear sequence of energy transfer
B. Network of interconnected food chains
C. Diagram showing only decomposers
D. Pyramid of energy
Q554. Which factor is density-dependent in population regulation?
A. Floods
B. Fires
C. Predation
D. Drought
Q555. The movement of individuals out of a population is called:
A. Immigration

B. Emigration
C. Dispersion
D. Infiltration

Q556. The layer of Earth where organisms live and interact is known as the:
A. Lithosphere
B. Hydrosphere
C. Atmosphere
D. Biosphere

Q557. Which of the following is an abiotic factor in an ecosystem?
A. Fungi
B. Bacteria
C. Temperature
D. Plants

Q558. A population crash occurs when:
A. Carrying capacity increases suddenly
B. Resource availability declines sharply
C. Population reaches stable equilibrium
D. Birth rate equals death rate

Q559. The zone of ocean closest to shore where sunlight penetrates is the:
A. Benthic zone
B. Intertidal zone
C. Pelagic zone
D. Abyssal zone

Q560. Which of the following best describes an energy pyramid?
A. Shows matter recycling
B. Illustrates population size changes
C. Represents energy transfer between trophic levels
D. Displays seasonal migration patterns

Q561. Which ecological principle states that no two species can occupy the same niche indefinitely?
A. Competitive exclusion principle
B. Law of tolerance
C. Hardy-Weinberg equilibrium
D. Ecological succession

Q562. In an aquatic ecosystem, the primary producers are usually:
A. Fish
B. Algae
C. Zooplankton
D. Crustaceans

Q563. The role and position of a species within its ecosystem is its:
A. Habitat
B. Niche
C. Territory
D. Population

Q564. The relationship where both species benefit is called:
A. Commensalism
B. Mutualism
C. Parasitism
D. Competition

Q565. A group of populations of different species living in the same area is called a:
A. Community
B. Ecosystem
C. Habitat
D. Biome
Q566. Which of the following gases is most directly linked to global warming?
A. Oxygen
B. Nitrogen
C. Carbon dioxide
D. Argon
Q567. Which aquatic zone receives little or no sunlight?
A. Euphotic zone
B. Aphotic zone
C. Littoral zone
D. Neritic zone
Q568. Population density is defined as:
A. Number of organisms per unit area
B. Birth rate of a species
C. Distribution of species globally
D. Migration pattern of individuals
Q569. The energy loss between trophic levels is mainly due to:
A. Photosynthesis
B. Respiration and heat loss
C. Decomposition
D. Migration
Q570. The removal of which species type often results in ecosystem collapse?
A. Keystone species
B. Primary consumer
C. Producer
D. Decomposer
Q571. Which organisms break down dead matter, returning nutrients to the soil?
A. Producers
B. Consumers
C. Decomposers
D. Herbivores
Q572. The part of the carbon cycle that removes CO_2 from the atmosphere is:
A. Combustion
B. Photosynthesis
C. Respiration
D. Decomposition
Q573. The maximum number of individuals an environment can support is called:
A. Birth rate
B. Carrying capacity
C. Density rate
D. Ecological niche
Q574. Which organisms occupy the lowest trophic level in an ecosystem?
A. Herbivores
B. Producers

C. Carnivores
D. Decomposers

Q575. The term "biodiversity" refers to:
A. The number of species in a habitat
B. The genetic variation within populations
C. The variety of life across ecosystems
D. All of the above

Q576. Which layer of Earth is composed primarily of solid iron and nickel?
A. Crust
B. Mantle
C. Outer core
D. Inner core

Q577. What evidence most strongly supports the theory of plate tectonics?
A. Similar fossils found on different continents
B. Distribution of deserts across the globe
C. Variations in atmospheric gases
D. Seasonal climate changes

Q578. Which process is primarily responsible for the formation of sedimentary rocks?
A. Melting
B. Cooling and crystallization
C. Compaction and cementation
D. Metamorphism

Q579. What drives the movement of tectonic plates?
A. Gravitational attraction of the Moon
B. Convection currents in the mantle
C. Solar radiation pressure
D. Differential rotation of the core

Q580. Which of the following best explains why oceanic crust is younger than continental crust?
A. Oceanic crust forms continually at mid-ocean ridges and is recycled in subduction zones.
B. Continental crust forms faster than oceanic crust.
C. Oceanic crust is less dense and therefore erodes quickly.
D. Continental crust forms from mantle plumes.

Q581. What is the primary cause of earthquakes along transform plate boundaries?
A. Compression
B. Shearing forces
C. Tension
D. Volcanic activity

Q582. Which rock type forms when magma cools and solidifies beneath Earth's surface?
A. Extrusive igneous rock
B. Intrusive igneous rock
C. Sedimentary rock
D. Metamorphic rock

Q583. Which process transforms sedimentary rocks into metamorphic rocks?
A. Weathering
B. Deposition
C. Heat and pressure
D. Melting

Q584. Which layer of the atmosphere contains the ozone layer?

A. Troposphere
B. Stratosphere
C. Mesosphere
D. Thermosphere

Q585. What geological feature typically forms at a divergent plate boundary?
A. Deep ocean trench
B. Mid-ocean ridge
C. Fold mountain range
D. Transform fault

Q586. Which process in the water cycle involves water vapor changing directly to liquid?
A. Evaporation
B. Transpiration
C. Condensation
D. Precipitation

Q587. What causes the Coriolis effect on Earth?
A. Tidal forces
B. Earth's rotation
C. Gravitational pull of the Moon
D. Differences in solar radiation

Q588. Which natural event is most responsible for the release of large amounts of sulfur dioxide into the atmosphere?
A. Earthquakes
B. Volcanic eruptions
C. Hurricanes
D. Glacial melting

Q589. What type of boundary produces mountain ranges like the Himalayas?
A. Divergent
B. Convergent continental-continental
C. Transform
D. Oceanic-oceanic convergent

Q590. Which process involves the breakdown of rocks through chemical reactions?
A. Physical weathering
B. Erosion
C. Chemical weathering
D. Deposition

Q591. Which component of the hydrosphere contains the majority of Earth's freshwater?
A. Rivers
B. Lakes
C. Glaciers and ice caps
D. Groundwater

Q592. Which process returns carbon from the atmosphere to the biosphere?
A. Combustion
B. Photosynthesis
C. Volcanic activity
D. Respiration

Q593. What type of rock provides the best evidence for Earth's past environments?
A. Igneous rock
B. Sedimentary rock

C. Metamorphic rock
D. Volcanic rock

Q594. Which factor contributes most to global sea-level rise?

A. Volcanic activity
B. Melting glaciers and ice sheets
C. Increased sediment deposition
D. Reduced rainfall

Q595. What is the main source of energy that drives the water cycle?

A. Earth's core heat
B. Solar radiation
C. Radioactive decay
D. Ocean currents

Q596. Which natural hazard is most directly associated with subduction zones?

A. Tornadoes
B. Earthquakes and volcanic eruptions
C. Flooding
D. Glacial retreat

Q597. Which process describes the circulation of water among Earth's systems?

A. Nitrogen cycle
B. Rock cycle
C. Water cycle
D. Carbon cycle

Q598. Which layer of Earth's interior is responsible for generating its magnetic field?

A. Crust
B. Mantle
C. Outer core
D. Inner core

Q599. Which geologic principle states that the present is the key to the past?

A. Catastrophism
B. Uniformitarianism
C. Superposition
D. Original horizontality

Q600. Which factor most influences the distribution of deserts on Earth?

A. Latitude and global wind patterns
B. Ocean depth
C. Volcanic activity
D. Earth's magnetic field

Answer keys and Explanations (501-600)

Q501. Answer: C

Explanation: Chloroplasts are double-membraned organelles containing chlorophyll pigments that capture sunlight for photosynthesis, converting light energy into glucose and oxygen. This process sustains nearly all life by producing organic matter and atmospheric oxygen. Unique to plants and some protists, chloroplasts evolved through endosymbiosis and illustrate eukaryotic adaptation to autotrophic energy conversion mechanisms.

Q502. Answer: A

Explanation: Mutations introduce new alleles, serving as the raw material for evolution. When advantageous variations arise, natural selection amplifies their frequency within the population, enabling adaptation to changing conditions. Although often neutral or harmful, beneficial mutations are crucial for species diversification and long-term evolutionary success through cumulative genetic change.

Q503. Answer: B

Explanation: Asexual reproduction, including binary fission and budding, produces offspring genetically identical to the parent since it involves no gamete fusion or genetic recombination. This mechanism allows rapid population growth and energy efficiency but limits genetic diversity, reducing adaptability under environmental stress compared to sexually reproducing organisms.

Q504. Answer: B

Explanation: All animals are multicellular, heterotrophic eukaryotes that ingest food for energy. Their lack of cell walls allows cellular flexibility and diverse body plans. This combination of heterotrophy and multicellularity supports complex tissue organization, movement, and adaptation, distinguishing animals from plants, fungi, and unicellular organisms.

Q505. Answer: B

Explanation: Meiosis reduces chromosome number by half, producing haploid gametes, and fertilization restores diploidy, ensuring genetic continuity. This alternation of reduction and restoration prevents chromosome duplication while introducing variability through recombination, balancing stability with diversity across generations in complex eukaryotic life cycles.

Q506. Answer: B

Explanation: Fossil records reveal transitional forms and gradual modifications in morphology over time, illustrating evolutionary change. These patterns align with descent with modification, as described by Darwin, corroborated by genetic, anatomical, and molecular data linking ancient and modern species through continuous adaptation and divergence.

Q507. Answer: D

Explanation: The species level defines organisms that can interbreed and produce fertile offspring, representing the most specific classification. It encapsulates unique genetic, behavioral, and morphological characteristics, providing precision in taxonomy and evolutionary analysis within the hierarchical system of biological classification.

Q508. Answer: B

Explanation: Commensalism is a symbiotic relationship where one species benefits without affecting the other. Examples include barnacles on whales or epiphytic plants on trees. This interaction demonstrates ecological balance, showing how species coexist and occupy diverse niches within ecosystems.

Q509. Answer: A

Explanation: Enzymes are specialized proteins that catalyze biochemical reactions by lowering activation energy, accelerating metabolic processes essential to life. Their specificity arises from precise amino acid sequences and folding, which create active sites for substrate binding and regulation, maintaining efficient cellular metabolism.

Q510. Answer: B

Explanation: During photosynthesis, light-dependent reactions occurring in the thylakoid membranes split water molecules (photolysis), releasing oxygen as a byproduct. This process converts solar energy into chemical energy (ATP and NADPH), driving the Calvin cycle, and replenishes atmospheric oxygen critical for aerobic life forms.

Q511. Answer: B

Explanation: Vascular plants have specialized conducting tissues—xylem for water and minerals, phloem for sugars—enabling larger size and terrestrial colonization. Nonvascular plants, like mosses, rely on diffusion, limiting size and habitat. This vascular adaptation marked a major evolutionary advancement in plant physiology and ecology.

Q512. Answer: C

Explanation: Sexual reproduction combines genetic material from two parents, producing offspring with unique genetic combinations. This variation underpins adaptation and evolution, contrasting with asexual methods that yield identical progeny. Gamete fusion promotes resilience against environmental change and genetic diseases.

Q513. Answer: B

Explanation: Ribosomes assemble amino acids into proteins using mRNA templates, essential for cellular structure, enzymes, and regulation. Found free in cytoplasm or attached to rough ER, ribosomes function universally, demonstrating the conserved molecular machinery shared by all life forms.

Q514. Answer: A

Explanation: Stabilizing selection maintains population stability by favoring average phenotypes and reducing extremes. This process preserves adaptation to consistent environments, common in traits like human birth weight, balancing survival advantages of intermediate characteristics across generations.

Q515. Answer: B

Explanation: DNA encodes hereditary instructions governing protein synthesis and organismal traits. Its sequence of nucleotides directs RNA transcription and subsequent translation into functional proteins, maintaining continuity of life and genetic inheritance across cellular and evolutionary timescales.

Q516. Answer: B

Explanation: Genetic drift causes random allele frequency shifts independent of fitness, particularly in small populations where chance events dominate. It can lead to loss of genetic diversity and fixation of neutral traits, influencing population structure and evolutionary trajectories.

Q517. Answer: B

Explanation: Desert plants exhibit xerophytic adaptations, including thick waxy cuticles reducing evaporation and CAM photosynthesis, where stomata open at night to conserve water. These physiological and structural modifications ensure survival under arid, high-temperature environments with limited moisture.

Q518. Answer: B

Explanation: Biodiversity encompasses diversity across species, genetic variation, and ecosystem complexity. It enhances resilience, stability, and ecosystem productivity. Conserving biodiversity preserves ecological balance, evolutionary potential, and essential resources for human and environmental sustainability.

Q519. Answer: C

Explanation: The excretory system, including kidneys and associated structures, filters blood and regulates water, ion, and waste balance, maintaining homeostasis. Through osmoregulation and excretion, this system sustains internal equilibrium critical to organismal survival across environments.

Q520. Answer: C

Explanation: Mitosis duplicates genetic material and divides a parent cell into two genetically identical daughter cells, driving growth, repair, and maintenance in multicellular organisms. Controlled mitotic cycles ensure tissue integrity and genetic stability across cell generations.

Q521. Answer: B

Explanation: Migration, a behavioral adaptation, enables animals to exploit favorable climates or resources seasonally, enhancing survival and reproduction. Such behaviors evolve through natural selection, aligning physiology and environment to optimize energy use and population persistence.

Q522. Answer: B

Explanation: Decomposers, such as fungi and bacteria, break down organic waste and dead organisms, returning nutrients to soil and water. This nutrient cycling sustains producers, closes energy loops, and ensures long-term ecosystem productivity and ecological equilibrium.

Q523. Answer: B

Explanation: Convergent evolution occurs when distinct species evolve analogous traits independently due to similar environmental pressures. Examples include wings in bats and insects. This phenomenon highlights natural selection's consistent influence on functional adaptations across diverse lineages.

Q524. Answer: B

Explanation: Lysosomes contain hydrolytic enzymes that degrade macromolecules, old organelles, and pathogens. This intracellular digestion maintains cellular health, preventing accumulation of harmful materials and supporting renewal processes essential to cellular metabolism and defense mechanisms.

Q525. Answer: B

Explanation: Genetic variation provides the foundation for natural selection by producing diverse traits that enhance survival under fluctuating environments. Populations with higher variation adapt more effectively, ensuring evolutionary continuity and long-term species persistence amid ecological pressures.

Q526. Answer: B

Explanation: An ecosystem includes all living organisms in a particular area interacting with each other and their nonliving environment. This interaction involves energy flow and nutrient cycling. Understanding ecosystems integrates biology and environmental science, emphasizing the dynamic balance between organisms and abiotic factors essential for sustainability, biodiversity conservation, and resilience of natural systems under environmental change.

Q527. Answer: B

Explanation: The sun provides energy for photosynthesis, which supports primary producers that form the foundation of food webs. This solar input drives biological and physical processes sustaining ecosystems. Understanding solar energy flow reveals the connection between photosynthesis, trophic levels, and global climate patterns, highlighting energy's central role in maintaining life and regulating ecological stability across the biosphere.

Q528. Answer: B

Explanation: Biotic factors include all living organisms within an ecosystem that interact with each other and with abiotic components. They influence energy flow, nutrient cycles, and community structure. Understanding biotic interactions—competition, predation, and symbiosis—clarifies how ecosystems maintain balance, supporting biodiversity and ecosystem services critical to human survival and environmental sustainability.

Q529. Answer: B

Explanation: Decomposition breaks down dead organic matter, releasing nutrients back into the soil and atmosphere for reuse by producers. Bacteria and fungi perform this crucial function. Understanding decomposition connects biology and chemistry, explaining nutrient cycling, soil fertility, and energy flow essential to ecosystem productivity and long-term sustainability in both terrestrial and aquatic systems globally.

Q530. Answer: A

Explanation: Primary consumers feed directly on producers, transferring energy from plants to higher trophic levels. They form the link between autotrophs and secondary consumers. Understanding trophic relationships clarifies energy flow and ecological pyramids, highlighting the efficiency limits and balance that sustain population sizes, nutrient distribution, and ecosystem stability across natural and managed environments.

Q531. Answer: B

Explanation: The nitrogen cycle converts atmospheric nitrogen into ammonia and nitrates through nitrogen fixation, enabling plants to incorporate it into biomolecules. Bacteria and decomposers regulate this transformation. Understanding the nitrogen cycle connects microbiology, chemistry, and ecology, emphasizing nutrient availability, agricultural productivity, and human impacts through fertilizers altering ecosystem balance and water quality globally.

Q532. Answer: C

Explanation: Tropical rainforests, with warm temperatures and abundant rainfall, support the highest species diversity and complex food webs. Their layered vegetation creates numerous niches. Understanding biodiversity patterns reveals ecological stability, productivity, and conservation priorities. Rainforests highlight how climate and resources influence life distribution, emphasizing the importance of preserving biodiversity hotspots for global ecological health.

Q533. Answer: A

Explanation: Carrying capacity represents the largest population an environment can support indefinitely without resource depletion. It is determined by food, habitat, and environmental resistance. Understanding this concept connects ecology and resource management, illustrating limits to growth and sustainability. It explains population dynamics, conservation strategies, and environmental balance critical to maintaining long-term ecosystem stability.

Q534. Answer: D

Explanation: Tertiary consumers occupy the highest trophic level, receiving the least energy due to losses at each transfer (about 10% efficiency). Understanding energy pyramids highlights constraints on food chain length and biomass. It links thermodynamics and ecology, emphasizing energy flow limits shaping ecosystem structure, predator-prey relationships, and environmental sustainability across biological communities globally.

Q535. Answer: A

Explanation: Deserts experience minimal rainfall and high evaporation, supporting drought-resistant plants and nocturnal animals. Adaptations like water storage and heat tolerance enable survival. Understanding desert ecology illustrates adaptations to abiotic stress, species interactions, and biogeographic distribution. It integrates climate science and evolutionary biology, showing life's resilience in extreme environments shaped by energy and resource scarcity.

Q536. Answer: A

Explanation: Temperature and precipitation patterns shape biome distribution and productivity by influencing soil type, vegetation, and biodiversity. Understanding these climatic variables links geography, ecology, and climate science, explaining global ecosystem patterns. They determine whether environments support forests, grasslands, or deserts, underscoring climate's role in ecological adaptation, resilience, and sustainability across the biosphere.

Q537. Answer: C

Explanation: The water cycle circulates water through evaporation, condensation, and precipitation, connecting oceans, land, and atmosphere. It supports climate regulation, nutrient transport, and ecosystem productivity. Understanding this cycle integrates hydrology, meteorology, and ecology, emphasizing water's critical role in sustaining life, weather patterns, and agricultural systems essential for environmental and human stability globally.

Q538. Answer: A

Explanation: Pioneer species, such as lichens and mosses, initiate ecosystem development on bare rock or disturbed soil, modifying conditions for later species. Understanding succession explains ecosystem recovery, resilience, and soil formation. It links disturbance ecology and time scales, illustrating how natural and human-impacted systems regenerate through orderly community replacement processes fundamental to ecological balance and restoration.

Q539. Answer: A

Explanation: Producers have greater biomass than consumers because energy diminishes at each trophic level due to metabolic losses. Understanding biomass distribution clarifies energy constraints and productivity patterns. It integrates physics and ecology, demonstrating how photosynthesis supports higher levels of life and how disruptions to producers cascade through food webs affecting overall ecosystem stability and resilience.

Q540. Answer: C

Explanation: Keystone species exert strong control on community composition relative to their abundance, maintaining ecosystem diversity and function. Examples include sea otters controlling urchin populations. Understanding keystone species connects ecology, conservation, and management, illustrating how removing such species destabilizes ecosystems. It emphasizes interconnectedness and balance, critical concepts in biodiversity preservation and ecological research worldwide.

Q541. Answer: B

Explanation: Commensalism is a symbiotic relationship where one species benefits without affecting the other. Examples include barnacles on whales. Understanding symbiotic relationships connects ecology, evolution, and species interactions. It reveals ecological cooperation and neutrality, emphasizing coexistence mechanisms within ecosystems essential for community diversity, balance, and long-term ecosystem sustainability.

Q542. Answer: C

Explanation: Carbon dioxide released from fossil fuel combustion traps heat in the atmosphere, enhancing the greenhouse effect and contributing to climate change. Understanding carbon cycling links human activity, atmospheric chemistry, and ecology. It underscores global environmental challenges requiring mitigation through sustainable energy practices and conservation to maintain ecological balance and prevent irreversible climate impacts.

Q543. Answer: D

Explanation: An energy pyramid shows energy transfer between trophic levels, highlighting losses as heat at each stage. It reflects energy's unidirectional flow from producers to consumers. Understanding energy pyramids integrates physics and ecology, emphasizing limits on trophic structure, resource availability, and ecosystem efficiency crucial for ecological modeling and conservation management of natural systems.

Q544. Answer: B

Explanation: Resource availability determines how many individuals an environment can sustain. Food, water, and space limit population growth and influence competition. Understanding resource limitation connects ecology, population biology, and sustainability, highlighting how resource management maintains balance in ecosystems and human societies by ensuring renewable use and ecological integrity under environmental constraints.

Q545. Answer: C

Explanation: Invasive species outcompete native organisms, disrupt ecosystems, and reduce biodiversity. Examples include zebra mussels and kudzu. Understanding biological invasions integrates ecology, evolution, and conservation management, emphasizing human roles in species translocation. It highlights ecosystem vulnerability and the need for preventive measures and restoration to maintain ecological balance and protect biodiversity globally.

Q546. Answer: B

Explanation: Photosynthesis uses light energy to convert carbon dioxide and water into glucose and oxygen, forming the basis of the carbon cycle. It supports all trophic levels by producing energy-rich compounds. Understanding photosynthesis integrates biology, chemistry, and ecology, demonstrating how primary production drives ecosystems and regulates atmospheric carbon balance essential to climate and life.

Q547. Answer: B

Explanation: Logistic growth describes population expansion that slows as it approaches carrying capacity, forming an S-shaped curve. It reflects resource limitations and density-dependent factors. Understanding logistic models connects mathematics, ecology, and environmental science, explaining how population regulation maintains ecosystem stability and informs wildlife management, conservation, and sustainable resource practices globally.

Q548. Answer: B

Explanation: Nitrogen-fixing bacteria, such as Rhizobium in legume root nodules, convert atmospheric nitrogen into ammonia usable by plants. This process sustains soil fertility and productivity. Understanding nitrogen fixation integrates microbiology, plant biology, and agriculture, emphasizing symbiosis and nutrient cycling fundamental to sustainable farming and ecosystem nutrient balance worldwide.

Q549. Answer: B

Explanation: r-strategists reproduce rapidly, producing many offspring with minimal parental care, exploiting unstable environments. Examples include insects and annual plants. Understanding reproductive strategies connects evolution, ecology, and life-history theory, showing trade-offs between quantity and quality of offspring that influence population dynamics, adaptability, and survival strategies across species and ecosystems.

Q550. Answer: D

Explanation: The biosphere encompasses all ecosystems where life exists, including land, water, and air. It integrates all living organisms and their interactions with abiotic factors. Understanding the biosphere links ecology, geology, and atmospheric science, emphasizing Earth's interconnected systems and the importance of conserving life-supporting processes vital for planetary stability and sustainability.

Q551. Answer: B

Explanation: Ecological succession is the gradual process of community change after disturbance, leading from pioneer species to a stable climax community. It involves soil formation, species replacement, and biomass increase. Understanding succession integrates time, ecology, and environmental recovery, illustrating how ecosystems regenerate naturally and maintain balance through resilience and species interactions essential to long-term ecosystem stability.

Q552. Answer: C

Explanation: The taiga, or boreal forest, is characterized by cold temperatures, evergreen conifers, and short growing seasons. It stores vast carbon reserves in soil and vegetation. Understanding taiga ecology links climate, geography, and global carbon cycling, emphasizing this biome's role in regulating Earth's climate and supporting cold-adapted species vital to northern hemisphere biodiversity.

Q553. Answer: B

Explanation: A food web shows complex feeding relationships linking multiple food chains within an ecosystem. It demonstrates how energy and nutrients flow among producers, consumers, and decomposers. Understanding food webs integrates ecology and systems theory, revealing interdependence and stability in ecosystems. It highlights resilience against species loss, emphasizing balance and biodiversity maintenance across ecological networks.

Q554. Answer: C

Explanation: Density-dependent factors, such as predation and disease, intensify as population size increases, regulating growth through biotic interactions. Understanding these mechanisms connects ecology, population biology, and conservation, explaining how natural checks and balances stabilize populations and prevent resource depletion, maintaining equilibrium within ecosystems under variable environmental pressures over time.

Q555. Answer: B

Explanation: Emigration refers to individuals leaving a population, reducing its size and affecting genetic diversity. It balances immigration in shaping population dynamics. Understanding movement patterns integrates population ecology and evolution, highlighting how dispersal influences gene flow, adaptation, and species distribution across ecosystems under environmental and ecological pressures over generations.

Q556. Answer: D

Explanation: The biosphere encompasses all ecosystems where living organisms interact with air, land, and water. It integrates biotic and abiotic processes that sustain life. Understanding the biosphere links ecology, geology, and climate systems, emphasizing interconnectedness and global balance critical for maintaining life-supporting environments essential to biodiversity and human survival.

Q557. Answer: C

Explanation: Abiotic factors are nonliving environmental components such as temperature, light, and water that influence living organisms. They determine habitat conditions and species distribution. Understanding abiotic influences connects ecology and physical geography, emphasizing environmental variability's role in shaping adaptation, energy flow, and ecosystem productivity across global biomes and ecological systems.

Q558. Answer: B

Explanation: A population crash follows resource depletion, disease, or predation, causing rapid population decline. Understanding population crashes integrates ecology, conservation, and resource management, emphasizing sustainability limits and ecosystem resilience. It demonstrates how environmental feedback regulates populations, preventing overexploitation and maintaining balance necessary for ecological health and biodiversity stability.

Q559. Answer: B

Explanation: The intertidal zone lies between high and low tide marks, receiving ample sunlight and experiencing strong wave action. Organisms here adapt to fluctuating conditions. Understanding marine zonation connects oceanography and ecology, revealing how light availability, salinity, and tides shape biodiversity patterns critical to coastal productivity and marine food webs.

Q560. Answer: C

Explanation: An energy pyramid illustrates energy flow from producers to top consumers, with diminishing energy at each level. It reflects ecological efficiency and trophic structure. Understanding energy pyramids links thermodynamics and ecology, emphasizing the limits of energy transfer and their influence on ecosystem organization, productivity, and conservation of energy within biological systems.

Q561. Answer: A

Explanation: The competitive exclusion principle states that species competing for identical resources cannot coexist; one will outcompete the other. Understanding this concept integrates ecology and evolution, explaining resource partitioning and specialization. It highlights biodiversity mechanisms and coexistence strategies that promote ecosystem balance and species adaptation in dynamic environmental systems globally.

Q562. Answer: B

Explanation: Algae, including phytoplankton, are photosynthetic producers that form the base of aquatic food webs. They convert solar energy into biomass, supporting higher trophic levels. Understanding aquatic productivity integrates biology, chemistry, and environmental science, emphasizing the importance of primary production for oxygen generation, nutrient cycling, and sustaining marine and freshwater ecosystems globally.

Q563. Answer: B

Explanation: A niche describes how a species interacts with its environment, including resource use, behavior, and ecological role. Understanding niches connects ecology, evolution, and behavior, explaining coexistence and specialization that reduce competition. It emphasizes adaptability and biodiversity maintenance, ensuring ecological balance through complementary roles among species within ecosystems.

Q564. Answer: B

Explanation: Mutualism is a symbiotic relationship in which both species benefit, such as bees pollinating flowers. Understanding mutualism integrates ecology and evolution, demonstrating interdependence and cooperation in nature. It highlights how positive interactions enhance survival, ecosystem function, and biodiversity, promoting stability and resilience in natural communities and environmental systems.

Q565. Answer: A

Explanation: A community includes all populations of different species coexisting and interacting within a specific area. It encompasses competition, predation, and symbiosis. Understanding communities connects ecology, behavior, and population dynamics, emphasizing interaction networks that maintain biodiversity, energy flow, and balance within ecosystems vital to environmental stability and resilience globally.

Q566. Answer: C

Explanation: Carbon dioxide, a greenhouse gas, traps infrared radiation, contributing to global temperature rise. Its levels have increased due to fossil fuel combustion and deforestation. Understanding greenhouse gases connects atmospheric science and ecology, emphasizing climate change impacts, feedback mechanisms, and the need for global mitigation strategies to preserve ecological and human systems.

Q567. Answer: B

Explanation: The aphotic zone lies below the reach of sunlight, preventing photosynthesis. Organisms here rely on chemosynthesis or detritus for energy. Understanding aquatic zonation integrates marine biology and environmental science, illustrating adaptations to darkness and pressure and highlighting deep-sea ecosystems' role in nutrient cycling and global biodiversity maintenance.

Q568. Answer: A

Explanation: Population density measures the number of individuals per unit area or volume, reflecting how populations use space and resources. Understanding density integrates ecology, demography, and management, informing conservation, resource allocation, and habitat planning. It reveals relationships between species, environmental capacity, and sustainability of natural and human-influenced systems.

Q569. Answer: B

Explanation: Energy decreases between trophic levels because organisms use most energy for respiration, growth, and maintenance, releasing heat. Only about 10% transfers upward. Understanding energy loss integrates biology and thermodynamics, explaining ecosystem efficiency limits and the importance of producers in sustaining food webs essential for ecological balance and productivity.

Q570. Answer: A

Explanation: Keystone species play critical roles in maintaining community structure and biodiversity. Their removal can cause trophic cascades and ecosystem collapse. Understanding keystone dynamics connects conservation, ecology, and evolution, emphasizing species interdependence and the importance of protecting pivotal organisms to preserve ecosystem function, stability, and resilience globally.

Q571. Answer: C

Explanation: Decomposers such as bacteria and fungi break down organic matter, recycling nutrients into the environment. They close nutrient cycles, sustaining primary production. Understanding decomposition integrates biology, chemistry, and ecology, emphasizing decomposition's essential role in soil fertility, ecosystem regeneration, and global nutrient and carbon cycling processes supporting life.

Q572. Answer: B

Explanation: Photosynthesis captures atmospheric CO_2, converting it into glucose through light-driven reactions in plants and algae. Understanding this process connects plant biology and atmospheric science, showing carbon's flow through living systems. It emphasizes photosynthesis' role in regulating climate, providing energy, and maintaining oxygen balance essential for all aerobic life.

Q573. Answer: B

Explanation: Carrying capacity defines the environment's sustainable population limit determined by resource availability and environmental resistance. Understanding carrying capacity connects ecology, sustainability, and conservation, providing insights into growth control and ecosystem management. It ensures balanced resource use while maintaining biodiversity and ecological integrity for long-term environmental health.

Q574. Answer: B

Explanation: Producers, such as plants and algae, form the foundation of food chains by converting solar energy into chemical energy through photosynthesis. They sustain all higher trophic levels. Understanding producer roles integrates ecology and biochemistry, emphasizing energy transfer, ecosystem productivity, and environmental balance necessary for maintaining life and supporting all consumer organisms.

Q575. Answer: D

Explanation: Biodiversity encompasses species richness, genetic diversity, and ecosystem variety, representing life's complexity on Earth. Understanding biodiversity integrates ecology, genetics, and conservation, emphasizing its role in resilience, stability, and ecosystem services. It highlights the urgency of protecting diverse habitats against human impact to preserve ecological functions and global environmental health.

Q576. Answer: D

Explanation: The inner core is solid and composed mostly of iron and nickel under immense pressure, despite extremely high temperatures. Its density and metallic composition generate Earth's magnetic field indirectly through interactions with the liquid outer core. This region represents Earth's deepest layer, influencing geodynamics, magnetic phenomena, and the thermal energy balance driving convection within the planet.

Q577. Answer: A

Explanation: Fossil correlations, such as Mesosaurus remains in South America and Africa, demonstrate continental movement over geological time. Combined with magnetic striping on ocean floors and earthquake patterns along plate boundaries, these provide compelling evidence that continents drift atop a dynamic lithosphere, reshaping Earth's surface through plate tectonics.

Q578. Answer: C

Explanation: Sedimentary rocks form from the compaction and cementation of sediments derived from preexisting rocks. Weathering and erosion transport these materials to depositional environments where layers accumulate. Over time, pressure and mineral precipitation bind particles into solid rock, recording Earth's surface conditions and biological history through stratified deposits.

Q579. Answer: B

Explanation: Mantle convection, caused by heat from radioactive decay and residual core energy, drives plate motion. Rising hot material at mid-ocean ridges and descending cooler slabs in subduction zones create circulation patterns. These dynamic forces continually reshape continents and ocean basins through processes like seafloor spreading and mountain building.

Q580. Answer: A

Explanation: New oceanic crust forms at mid-ocean ridges through volcanic activity, then moves outward and subducts into trenches, recycling within the mantle. This continuous creation-destruction cycle explains why oceanic crust averages only 200 million years old, while continental crust persists for billions, forming stable landmasses.

Q581. Answer: B

Explanation: Transform boundaries involve lateral sliding of plates past one another, generating intense shear stress. When accumulated strain exceeds rock strength, it releases suddenly as seismic waves, causing earthquakes. The San Andreas Fault exemplifies this mechanism, illustrating energy buildup and release during crustal movement along fault lines.

Q582. Answer: B

Explanation: Intrusive igneous rocks crystallize slowly beneath Earth's surface, allowing large mineral grains to form. Examples include granite and diorite. Their coarse textures reveal slow cooling within magma chambers, providing insights into subsurface geological processes and crustal formation mechanisms.

Q583. Answer: C

Explanation: Metamorphism occurs when sedimentary or igneous rocks undergo changes under elevated heat and pressure without melting. This alters mineral composition and texture, forming new structures such as schist or gneiss. These transformations reflect tectonic activity and reveal conditions deep within Earth's crust.

Q584. Answer: B

Explanation: The ozone layer, located within the stratosphere, absorbs harmful ultraviolet radiation, shielding life on Earth. Its stability depends on chemical balance between ozone formation and destruction. Variations in ozone concentration influence climate and biological health, making it critical for environmental protection and atmospheric research.

Q585. Answer: B

Explanation: Divergent boundaries, where plates move apart, allow magma to rise and solidify, forming mid-ocean ridges. This continuous volcanic activity generates new crust, exemplified by the Mid-Atlantic Ridge. Divergence drives seafloor spreading and influences ocean basin expansion and global tectonic cycles.

Q586. Answer: C

Explanation: Condensation converts water vapor into liquid droplets, forming clouds and fog when air cools to its dew point. This process releases latent heat, driving weather patterns and precipitation. Condensation is essential for atmospheric circulation and water cycle continuity, linking evaporation and rainfall.

Q587. Answer: B

Explanation: The Coriolis effect results from Earth's rotation, deflecting moving air and water to the right in the Northern Hemisphere and to the left in the Southern Hemisphere. This deflection shapes global wind and ocean current patterns, influencing weather systems and climate distribution across latitudes.

Q588. Answer: B

Explanation: Volcanic eruptions emit sulfur dioxide, which reacts with atmospheric water vapor to form sulfate aerosols. These particles reflect sunlight, potentially cooling Earth's surface. Large eruptions, such as Mount Pinatubo in 1991, demonstrate volcanic influence on short-term climate variation and atmospheric chemistry.

Q589. Answer: B

Explanation: When two continental plates collide at a convergent boundary, neither subducts due to similar densities. Instead, crustal compression uplifts rock layers, forming extensive mountain ranges like the Himalayas. This tectonic collision illustrates crustal deformation and ongoing geological evolution over millions of years.

Q590. Answer: C

Explanation: Chemical weathering alters minerals through reactions with water, oxygen, and acids, transforming rock composition. Processes like oxidation and hydrolysis produce secondary minerals and release ions into soil and water. This mechanism drives soil formation and influences landscape evolution over geological timescales.

Q591. Answer: C

Explanation: Approximately 68% of Earth's freshwater is locked in glaciers and polar ice caps. This frozen reserve influences sea level and climate through melting and accumulation cycles. Monitoring cryospheric changes provides critical insight into global warming and long-term water resource availability.

Q592. Answer: B

Explanation: Photosynthesis removes atmospheric carbon dioxide, converting it into organic compounds within plants and algae. This biochemical process sustains life by producing oxygen and forms the foundation of the carbon cycle, linking atmosphere, biosphere, and hydrosphere through continuous energy and matter exchange.

Q593. Answer: B

Explanation: Sedimentary rocks preserve fossils, textures, and chemical signatures of ancient environments. Layers reveal depositional conditions, such as riverbeds, oceans, or deserts, providing chronological records of Earth's history. These archives are vital for reconstructing past climates and biological evolution.

Q594. Answer: B

Explanation: Rising global temperatures accelerate melting of glaciers and polar ice sheets, adding water to oceans and increasing sea level. Thermal expansion of seawater further amplifies the effect. This dual process drives coastal flooding, habitat loss, and climate feedback impacting global ecosystems.

Q595. Answer: B

Explanation: Solar energy powers the water cycle by heating Earth's surface, causing evaporation and transpiration. Condensation and precipitation return water to the surface, completing the cycle. This continuous solar-driven process regulates temperature, distributes energy, and sustains weather and climate systems globally.

Q596. Answer: B

Explanation: Subduction zones, where oceanic plates descend beneath continental crust, produce intense earthquakes and volcanic activity. Melting of subducted material generates magma, fueling volcanic arcs. These regions mark the planet's most geologically active and hazardous areas due to tectonic interactions and crustal stress.

Q597. Answer: C

Explanation: The water cycle, or hydrologic cycle, circulates water through evaporation, condensation, precipitation, and runoff. It connects the atmosphere, hydrosphere, biosphere, and lithosphere, balancing energy and matter. This process regulates climate, supports life, and redistributes Earth's freshwater resources.

Q598. Answer: C

Explanation: The liquid outer core, composed mainly of molten iron and nickel, generates Earth's magnetic field through convective motion and the geodynamo effect. This magnetic field protects the planet from solar radiation, guides navigation, and preserves atmospheric integrity over geological time.

Q599. Answer: B

Explanation: Uniformitarianism, proposed by James Hutton, asserts that geological processes operating today, such as erosion and sedimentation, have acted similarly throughout Earth's history. This principle provides the foundation for interpreting ancient rock formations and reconstructing past environmental conditions.

Q600. Answer: A

Explanation: Deserts typically form around 30° latitude due to descending dry air in global circulation cells. High-pressure zones suppress cloud formation and precipitation. These patterns, driven by solar heating and rotation, explain the predictable placement of major desert regions worldwide.

Questions (601-700)

Q601. Which type of plate boundary is associated with seafloor spreading?

A. Convergent boundary

B. Transform boundary

C. Divergent boundary

D. Subduction zone

Q602. Which process is responsible for shaping U-shaped valleys in mountainous regions?

A. River erosion

B. Wind erosion

C. Glacial erosion

D. Volcanic activity

Q603. Which term describes the point within Earth where an earthquake originates?

A. Epicenter

B. Focus (hypocenter)

C. Fault line

D. Seismic node

Q604. Which method is most commonly used to determine the absolute age of rocks?

A. Stratigraphic correlation

B. Fossil succession

C. Radiometric dating

D. Cross-cutting relationships

Q605. Which layer of Earth contains the asthenosphere?

A. Crust

B. Outer core

C. Mantle

D. Inner core

Q606. Which type of weathering results from repeated freezing and thawing of water in rock cracks?

A. Oxidation

B. Frost wedging

C. Hydrolysis

D. Carbonation

Q607. Which statement best describes the principle of superposition?

A. Younger layers lie below older layers.

B. Older layers lie beneath younger layers.

C. Faults are always younger than folded rocks.

D. Sediment layers form randomly.

Q608. Which factor most influences the rate of chemical weathering?

A. Altitude

B. Rock hardness

C. Climate temperature and moisture

D. Earthquake activity

Q609. Which natural process is primarily responsible for the formation of sand dunes?

A. Glacial transport

B. Wind deposition

C. River flooding

D. Volcanic activity

Q610. Which part of the atmosphere is closest to Earth's surface?

A. Stratosphere
B. Troposphere
C. Mesosphere
D. Thermosphere

Q611. What is the main driver of deep ocean currents?
A. Wind patterns
B. Temperature and salinity differences
C. Tidal forces
D. Earth's rotation

Q612. Which type of volcanic eruption produces broad, gently sloping landforms?
A. Stratovolcanic eruption
B. Shield volcanic eruption
C. Cinder cone eruption
D. Caldera collapse

Q613. Which statement correctly describes groundwater?
A. It exists only in underground rivers.
B. It fills spaces between soil and rock particles.
C. It flows exclusively through volcanic rocks.
D. It is not part of the water cycle.

Q614. What is the boundary between two different air masses called?
A. Jet stream
B. Front
C. Convection cell
D. Pressure ridge

Q615. Which layer of Earth's atmosphere protects life from harmful ultraviolet radiation?
A. Troposphere
B. Stratosphere
C. Mesosphere
D. Exosphere

Q616. Which natural phenomenon best demonstrates the process of convection?
A. Ocean tides
B. Earthquakes
C. Mantle currents driving plate motion
D. Weathering of rocks

Q617. Which type of rock forms directly from magma cooling on Earth's surface?
A. Basalt
B. Granite
C. Schist
D. Limestone

Q618. Which process describes the downward movement of rock and soil under gravity?
A. Deposition
B. Mass wasting
C. Erosion
D. Weathering

Q619. Which atmospheric gas plays the largest role in the greenhouse effect?
A. Oxygen
B. Carbon dioxide
C. Nitrogen

D. Argon

Q620. Which event marked the transition from the Precambrian to the Paleozoic era?

A. Mass extinction of dinosaurs
B. Cambrian explosion of life
C. Formation of Earth's oceans
D. Ice age expansion

Q621. What phenomenon results from differences in atmospheric pressure across Earth's surface?

A. Tides
B. Winds
C. Earthquakes
D. Evaporation

Q622. Which type of boundary is associated with deep ocean trenches?

A. Divergent
B. Convergent
C. Transform
D. Passive margin

Q623. Which geologic law states that faults and intrusions are younger than the rocks they cut through?

A. Superposition
B. Original horizontality
C. Cross-cutting relationships
D. Lateral continuity

Q624. What causes the alternating magnetic patterns found on the ocean floor?

A. Tidal forces
B. Magnetic reversals recorded in basalt
C. Seismic activity
D. Sediment layering

Q625. Which human activity contributes most significantly to soil erosion?

A. Sustainable farming
B. Deforestation
C. Reforestation
D. Urban greening

Q626. Which model of the solar system places the Sun at the center?

A. Geocentric model
B. Heliocentric model
C. Ptolemaic system
D. Aristotelian system

Q627. The planets travel in what type of orbits around the Sun?

A. Circular
B. Elliptical
C. Parabolic
D. Hyperbolic

Q628. Which planet has the largest number of known moons?

A. Earth
B. Jupiter
C. Mars
D. Saturn

Q629. The asteroid belt lies between the orbits of:

A. Earth and Mars
B. Mars and Jupiter
C. Jupiter and Saturn
D. Venus and Earth

Q630. Which planet has the longest orbital period around the Sun?
A. Uranus
B. Neptune
C. Jupiter
D. Saturn

Q631. What causes the phases of the Moon?
A. Earth's shadow on the Moon
B. Changing positions of the Moon, Earth, and Sun
C. Rotation of the Moon on its axis
D. Atmospheric refraction

Q632. Which type of eclipse occurs when the Moon passes directly between the Sun and Earth?
A. Lunar eclipse
B. Solar eclipse
C. Partial lunar eclipse
D. Penumbral eclipse

Q633. The Sun's energy is produced primarily through:
A. Combustion
B. Nuclear fission
C. Nuclear fusion
D. Gravitational compression

Q634. The planet known for its prominent ring system is:
A. Neptune
B. Saturn
C. Uranus
D. Jupiter

Q635. What is the main reason Pluto is no longer classified as a major planet?
A. Its orbit crosses Neptune's
B. It lacks a moon
C. It does not clear its orbital neighborhood
D. It rotates too slowly

Q636. Which planet rotates on its axis in the opposite direction of most planets?
A. Venus
B. Mars
C. Jupiter
D. Neptune

Q637. The Sun's outermost layer visible during a total solar eclipse is called the:
A. Photosphere
B. Chromosphere
C. Corona
D. Core

Q638. What determines the color of a star?
A. Its age
B. Its temperature
C. Its composition

D. Its distance from Earth

Q639. What galaxy do we live in?

A. Andromeda Galaxy
B. Milky Way Galaxy
C. Triangulum Galaxy
D. Large Magellanic Cloud

Q640. What is a light-year a measure of?

A. Time
B. Brightness
C. Distance
D. Velocity

Q641. What is the name of the nearest star to Earth other than the Sun?

A. Sirius
B. Proxima Centauri
C. Betelgeuse
D. Vega

Q642. The Milky Way Galaxy is classified as what type of galaxy?

A. Elliptical
B. Spiral
C. Irregular
D. Lenticular

Q643. The Big Bang theory explains:

A. The creation of stars
B. The formation of planets
C. The origin and expansion of the universe
D. The appearance of galaxies only

Q644. Which instrument separates light into its component wavelengths for analysis?

A. Telescope
B. Microscope
C. Spectroscope
D. Barometer

Q645. Which law describes the relationship between a planet's orbital period and its distance from the Sun?

A. Newton's law of gravity
B. Kepler's third law
C. Einstein's relativity
D. Copernicus' principle

Q646. The apparent shift in position of a nearby star due to Earth's orbital motion is called:

A. Redshift
B. Parallax
C. Reflection
D. Refraction

Q647. What is the term for the remains of a massive star after a supernova explosion?

A. Neutron star
B. Red giant
C. White dwarf
D. Brown dwarf

Q648. The expansion of the universe is supported by evidence from:

A. Stellar parallax
B. Redshift of galaxies
C. Planetary retrograde motion
D. Solar wind

Q649. Which planet has a day longer than its year?

A. Mercury
B. Venus
C. Mars
D. Neptune

Q650. What is the approximate age of the universe according to current scientific understanding?

A. 6 billion years
B. 10 billion years
C. 13.8 billion years
D. 20 billion years

Q651. What is the primary component of the Sun?

A. Helium
B. Hydrogen
C. Carbon
D. Oxygen

Q652. Which planet is known as the "Red Planet" due to iron oxide on its surface?

A. Mercury
B. Venus
C. Mars
D. Jupiter

Q653. What type of celestial object is the Andromeda Galaxy?

A. Elliptical galaxy
B. Spiral galaxy
C. Irregular galaxy
D. Globular cluster

Q654. Which planet has the strongest magnetic field in the solar system?

A. Earth
B. Saturn
C. Jupiter
D. Neptune

Q655. Which type of galaxy contains little gas and dust and few new stars forming?

A. Spiral
B. Elliptical
C. Irregular
D. Barred spiral

Q656. What is the approximate average distance from the Earth to the Sun?

A. 93,000 miles
B. 93 million miles
C. 9.3 billion miles
D. 39 million miles

Q657. What type of star is the Sun classified as?

A. Red giant
B. White dwarf
C. Main-sequence star

D. Neutron star

Q658. The path a planet takes around the Sun is called its:

A. Axis
B. Orbit
C. Revolution
D. Rotation

Q659. Which gas is most abundant in the atmosphere of Jupiter?

A. Carbon dioxide
B. Nitrogen
C. Hydrogen
D. Methane

Q660. What type of celestial body is Halley's Comet?

A. Asteroid
B. Dwarf planet
C. Comet
D. Meteoroid

Q661. What causes the apparent retrograde motion of planets?

A. Their elliptical orbits
B. Earth's faster orbital speed
C. Axial tilt variations
D. Gravitational perturbations

Q662. Which planet has a day nearly equal in length to Earth's?

A. Mars
B. Mercury
C. Venus
D. Uranus

Q663. Which layer of the Sun is responsible for emitting visible light?

A. Core
B. Photosphere
C. Chromosphere
D. Corona

Q664. Which planet's rotation axis is tilted nearly 98 degrees, causing extreme seasons?

A. Saturn
B. Neptune
C. Uranus
D. Venus

Q665. Which telescope uses mirrors instead of lenses to collect light?

A. Refracting telescope
B. Reflecting telescope
C. Radio telescope
D. Spectrometer

Q666. The measure of a star's true brightness is called its:

A. Apparent magnitude
B. Absolute magnitude
C. Luminosity index
D. Radiant intensity

Q667. What is a supernova?

A. A newly formed star

B. A dying massive star's explosive end
C. A small red star's collapse
D. A comet impact event

Q668. Which planet has the most extreme temperature difference between day and night?
A. Venus
B. Mercury
C. Earth
D. Neptune

Q669. What phenomenon explains the change in frequency of light from moving celestial objects?
A. Doppler effect
B. Reflection
C. Refraction
D. Parallax

Q670. The boundary around a black hole beyond which nothing can escape is called the:
A. Singularity
B. Accretion disk
C. Event horizon
D. Nebular limit

Q671. Which galaxy is on a collision course with the Milky Way?
A. Triangulum Galaxy
B. Andromeda Galaxy
C. Small Magellanic Cloud
D. Whirlpool Galaxy

Q672. Which planet's atmosphere is primarily composed of carbon dioxide and thick clouds of sulfuric acid?
A. Venus
B. Earth
C. Mars
D. Neptune

Q673. What is the name for the flattened, rotating disk of gas and dust from which planets form?
A. Asteroid belt
B. Solar nebula
C. Kuiper belt
D. Oort cloud

Q674. The temperature of a star can be determined by its:
A. Size
B. Distance
C. Color
D. Mass

Q675. Which cosmic background radiation provides evidence of the Big Bang?
A. Infrared radiation
B. Cosmic microwave background radiation
C. Gamma-ray bursts
D. Ultraviolet radiation

Q676. Which of the following best describes the scientific method?
A. A fixed series of steps scientists must follow
B. A flexible process of observation, hypothesis, and testing
C. A single experiment to prove a theory

D. A way to confirm personal beliefs

Q677. Who is credited with formulating the laws of motion and universal gravitation?

A. Galileo Galilei

B. Isaac Newton

C. Johannes Kepler

D. Albert Einstein

Q678. The idea that scientific knowledge evolves through paradigm shifts was proposed by:

A. Francis Bacon

B. Karl Popper

C. Thomas Kuhn

D. Aristotle

Q679. Which of the following is essential to ensure reliability in scientific measurement?

A. Random guessing

B. Consistent procedures and calibration

C. Researcher intuition

D. Variable interpretation

Q680. The term "falsifiability" in science refers to:

A. The ability to confirm a hypothesis

B. The possibility that a claim can be proven false

C. The reproducibility of results

D. The ethical review process

Q681. Ethical standards in scientific research primarily exist to:

A. Encourage data manipulation

B. Protect researchers' reputations

C. Ensure integrity and protect participants

D. Speed up experimentation

Q682. The process of peer review serves to:

A. Promote competition between scientists

B. Ensure research meets professional standards

C. Hide flaws in research

D. Speed publication without evaluation

Q683. Which scientist's work laid the foundation for modern experimental physics by emphasizing systematic observation?

A. Galileo Galilei

B. Aristotle

C. René Descartes

D. Francis Crick

Q684. What distinguishes a scientific theory from a hypothesis?

A. A theory is an untested guess

B. A theory is broader and supported by evidence

C. A theory cannot change over time

D. A theory is purely speculative

Q685. The reproducibility of experiments is crucial because it:

A. Prevents new discoveries

B. Validates findings and builds credibility

C. Discourages collaboration

D. Reduces data transparency

Q686. Which of the following best defines empirical evidence?

A. Opinions from experts
B. Observations and data obtained through experience
C. Assumptions based on logic
D. Philosophical speculation

Q687. The shift from Newtonian mechanics to Einstein's relativity exemplifies:
A. A failure of the scientific method
B. A paradigm shift in scientific understanding
C. A regression to older theories
D. An error in observation

Q688. Which measurement property ensures that an assessment measures what it intends to?
A. Validity
B. Reliability
C. Objectivity
D. Variability

Q689. What is the main function of models in science?
A. To prove hypotheses beyond doubt
B. To simplify and represent complex systems
C. To eliminate uncertainty completely
D. To replace direct observation

Q690. The use of statistics in scientific analysis primarily helps to:
A. Increase measurement error
B. Draw objective conclusions from data
C. Reduce experiment size
D. Replace qualitative evidence

Q691. The principle of Occam's Razor suggests that:
A. Complex explanations are preferable
B. All hypotheses are equally valid
C. The simplest explanation consistent with evidence is best
D. Theories must remain untested

Q692. What role does skepticism play in scientific inquiry?
A. It prevents progress
B. It encourages critical evaluation of claims
C. It dismisses all new ideas
D. It replaces experimentation

Q693. The scientific revolution in the 17th century was characterized by:
A. A rejection of observation and experiment
B. Increased reliance on empirical evidence and mathematics
C. The dominance of superstition in science
D. The abandonment of reasoning

Q694. Double-blind experiments are used to:
A. Increase observer bias
B. Reduce both researcher and participant bias
C. Speed up testing
D. Eliminate control groups

Q695. Measurement error can be minimized by:
A. Using inconsistent instruments
B. Calibrating tools and standardizing procedures
C. Ignoring variability

D. Reducing sample size

Q696. Which factor most contributes to the credibility of scientific research findings?

A. Public opinion
B. Media coverage
C. Researcher reputation
D. Peer review and replication

Q697. The main difference between qualitative and quantitative research is:

A. Qualitative focuses on numerical data, quantitative on words
B. Quantitative uses numbers, qualitative explores meanings and contexts
C. Both rely solely on statistics
D. Quantitative ignores measurement precision

Q698. The replication crisis in science refers to:

A. Scientists refusing to share data
B. Difficulty reproducing previous research findings
C. Excessive publication speed
D. Overreliance on qualitative methods

Q699. Which concept emphasizes the moral responsibility of scientists toward society?

A. Objectivity
B. Social accountability
C. Theoretical neutrality
D. Reductionism

Q700. Why is open access publishing important in modern science?

A. It restricts access to verified information
B. It eliminates data transparency
C. It limits peer review opportunities
D. It increases public and academic availability of research

Answer keys and Explanations (601-700)

Q601. Answer: C

Explanation: Seafloor spreading occurs at divergent boundaries where tectonic plates move apart, allowing magma to rise and form new oceanic crust. As molten material cools, it solidifies, pushing older crust away. This process continuously renews ocean basins and provides key evidence for plate tectonics and continental drift theories, as observed at mid-ocean ridges such as the Mid-Atlantic Ridge.

Q602. Answer: C

Explanation: Glacial erosion occurs as moving ice grinds and carves through bedrock, transforming V-shaped river valleys into U-shaped glacial valleys. The immense pressure and weight of glaciers reshape landscapes, depositing moraines and striations that record past glacial movement and climate fluctuations over geological time.

Q603. Answer: B

Explanation: The focus, or hypocenter, is the initial rupture point where stored stress is released along a fault, generating seismic waves. The epicenter lies directly above it on Earth's surface. Understanding this distinction is crucial in seismology for assessing earthquake intensity, depth, and hazard distribution across fault systems.

Q604. Answer: C

Explanation: Radiometric dating measures the decay of radioactive isotopes, such as uranium to lead, to calculate absolute rock ages. This technique provides precise timelines for geological events, complementing relative dating methods. By analyzing isotope ratios, scientists establish Earth's history from mineral crystallization to continental evolution.

Q605. Answer: C

Explanation: The asthenosphere lies within the upper mantle, composed of partially molten rock that flows slowly beneath the rigid lithosphere. This ductile zone enables plate motion and deformation, acting as the mechanical foundation for tectonic activity and mantle convection processes shaping Earth's dynamic crust.

Q606. Answer: B

Explanation: Frost wedging occurs when water enters rock fractures, freezes, and expands, exerting pressure that widens cracks. Over time, this mechanical weathering fragments rock into smaller pieces, significantly shaping cold-climate landscapes. It demonstrates how simple physical processes can drive substantial geological transformation.

Q607. Answer: B

Explanation: The principle of superposition states that in undisturbed sedimentary sequences, the oldest layers are at the bottom and younger layers accumulate above them. This fundamental law of stratigraphy allows geologists to interpret relative ages and reconstruct depositional histories in sedimentary rock formations worldwide.

Q608. Answer: C

Explanation: Chemical weathering accelerates in warm, moist climates where water and heat promote reactions such as hydrolysis and oxidation. Tropical regions experience rapid decomposition of minerals, forming thick soils, whereas arid environments show slower alteration rates. Thus, climate profoundly governs Earth's surface chemistry and rock breakdown.

Q609. Answer: B

Explanation: Sand dunes form when wind transports and deposits loose sediment in arid or coastal regions. As wind velocity decreases, sand grains accumulate, creating various dune shapes depending on wind direction and sediment supply. Dune formation illustrates aeolian processes shaping Earth's landscapes through erosion and deposition.

Q610. Answer: B

Explanation: The troposphere is the lowest atmospheric layer, extending up to about 12 kilometers. It contains most of Earth's weather, clouds, and water vapor. Temperature decreases with altitude here, driving convection currents that produce storms, wind, and global weather patterns vital to Earth's climate system.

Q611. Answer: B

Explanation: Deep ocean currents, or thermohaline circulation, arise from differences in water density caused by temperature and salinity variations. Cold, salty water sinks at high latitudes, flowing along the ocean floor and redistributing heat globally. This conveyor system regulates climate and nutrient distribution across marine ecosystems.

Q612. Answer: B

Explanation: Shield volcanoes form from fluid basaltic lava that spreads over wide areas, producing low-angle slopes. Examples like Mauna Loa demonstrate how non-explosive eruptions generate extensive volcanic landscapes. Their structure contrasts sharply with steep, explosive stratovolcanoes, highlighting magma viscosity's role in eruption style.

Q613. Answer: B

Explanation: Groundwater occupies pores and fractures within soil and rock layers, forming aquifers that supply drinking and irrigation water. It moves slowly under gravity, replenished by precipitation infiltration. Understanding groundwater dynamics is essential for sustainable resource management and pollution prevention in hydrology.

Q614. Answer: B

Explanation: A front marks the transition between contrasting air masses with different temperatures and humidity. Interactions at fronts cause cloud formation, storms, and precipitation. Cold, warm, and stationary fronts drive mid-latitude weather systems, making them key features in meteorology and climate forecasting.

Q615. Answer: B

Explanation: The stratosphere contains the ozone layer, which absorbs ultraviolet radiation, preventing DNA damage in living organisms. Ozone balance depends on photochemical reactions involving oxygen molecules. Disruption by pollutants like CFCs highlights the interconnectedness of atmospheric chemistry and global environmental protection.

Q616. Answer: C

Explanation: Mantle convection occurs as hot, buoyant material rises and cool material sinks within Earth's mantle, transferring heat outward. This process drives tectonic plate movement and underlies volcanic and mountain-building activity. Convection exemplifies how internal energy maintains Earth's dynamic surface and geological renewal.

Q617. Answer: A

Explanation: Basalt forms when lava cools rapidly on Earth's surface, producing fine-grained crystals. It is the most common volcanic rock, composing much of the oceanic crust. Basaltic flows provide insights into volcanic processes and contribute to the continuous renewal of Earth's outer shell through seafloor spreading.

Q618. Answer: B

Explanation: Mass wasting, or slope movement, involves gravity-driven downslope transport of material without water or wind. Landslides, rockfalls, and slumps exemplify this process, which reshapes landscapes and poses natural hazards. It illustrates gravitational influence on surface stability and geological evolution.

Q619. Answer: B

Explanation: Carbon dioxide traps heat by absorbing infrared radiation emitted from Earth's surface, maintaining the planet's temperature balance. Human activities, such as burning fossil fuels, increase CO_2 levels, intensifying the greenhouse effect and driving global climate change with widespread environmental consequences.

Q620. Answer: B

Explanation: The Cambrian explosion, about 541 million years ago, marked a rapid diversification of multicellular life, ending the Precambrian. Fossil evidence reveals the emergence of most major animal phyla, signaling ecological complexity. This evolutionary milestone transformed marine ecosystems and defined the Paleozoic's beginning.

Q621. Answer: B

Explanation: Wind arises when air moves from high-pressure to low-pressure regions due to unequal solar heating. Pressure gradients drive circulation patterns that distribute heat and moisture globally. Understanding these dynamics is fundamental to meteorology and climate modeling, explaining weather variations and storm formation.

Q622. Answer: B

Explanation: Deep ocean trenches form at convergent boundaries where an oceanic plate subducts beneath another plate. These trenches, like the Mariana Trench, mark zones of intense pressure, volcanic activity, and seismic events, demonstrating destructive tectonic interactions that recycle Earth's lithosphere into the mantle.

Q623. Answer: C

Explanation: The principle of cross-cutting relationships asserts that any fault or intrusion disrupting existing strata must be younger than the layers it affects. This rule aids in relative dating and reconstructing geological events, allowing scientists to decipher the chronological order of rock formation and deformation.

Q624. Answer: B

Explanation: As magma at mid-ocean ridges solidifies, iron-rich minerals align with Earth's magnetic field. Periodic polarity reversals create symmetrical magnetic stripes on either side of ridges, recording seafloor spreading. These patterns provided critical evidence confirming plate tectonic theory and oceanic crust formation processes.

Q625. Answer: B

Explanation: Deforestation removes vegetation that stabilizes soil with roots, increasing erosion by wind and water. Exposed soil loses nutrients and structure, degrading land fertility and promoting sediment runoff. Reforestation and conservation practices mitigate these impacts, promoting sustainable ecosystem management and soil preservation.

Q626. Answer: B

Explanation: The heliocentric model, proposed by Copernicus, places the Sun at the center of the solar system with planets orbiting it. This concept revolutionized astronomy, replacing the Earth-centered view. Understanding the heliocentric model integrates physics, mathematics, and historical science, illustrating how observation and reasoning advanced celestial mechanics and redefined humanity's understanding of its place in the universe.

Q627. Answer: B

Explanation: Johannes Kepler determined that planets orbit the Sun in ellipses, with the Sun at one focus. This discovery refined planetary motion laws and replaced earlier circular assumptions. Understanding elliptical orbits combines observation and mathematics, showing gravitational influences and forming the foundation for Newton's law of universal gravitation explaining motion across cosmic scales.

Q628. Answer: D

Explanation: Saturn currently holds the record for the most confirmed moons, exceeding 140. Its gravitational field captures small objects, forming an extensive satellite system. Understanding Saturn's moons links planetary science, astrophysics, and orbital dynamics, revealing solar system diversity and how gravitational interactions shape planetary systems over billions of years of cosmic evolution.

Q629. Answer: B

Explanation: The asteroid belt, located between Mars and Jupiter, contains rocky remnants from the solar system's formation. Jupiter's gravity prevented their coalescence into a planet. Understanding the asteroid belt connects planetary formation, gravitational influence, and solar system evolution, offering insights into early materials that formed planets and the processes shaping celestial body distribution.

Q630. Answer: B

Explanation: Neptune, the farthest major planet from the Sun, takes about 165 Earth years to complete one orbit. Its distance results in slow motion and low solar energy. Understanding orbital periods connects Kepler's laws and celestial mechanics, showing how distance and gravity determine planetary motion and contribute to the solar system's dynamic structure.

Q631. Answer: B

Explanation: The Moon's phases occur as its illuminated portion changes relative to Earth and the Sun during orbit. Understanding lunar phases links geometry, light reflection, and observation, explaining why the Moon's shape appears to change cyclically. This concept reinforces spatial reasoning and celestial motion fundamentals central to astronomy and Earth–Moon–Sun relationships.

Q632. Answer: B

Explanation: A solar eclipse happens when the Moon blocks sunlight from reaching part of Earth, casting a shadow. Total, partial, or annular eclipses depend on alignment and distance. Understanding eclipses integrates celestial mechanics and geometry, illustrating orbital relationships, motion predictability, and how observation validates astronomical principles of alignment and timing within the solar system.

Q633. Answer: C

Explanation: The Sun's core converts hydrogen into helium through nuclear fusion, releasing vast energy that powers the solar system. Understanding fusion connects nuclear physics, thermodynamics, and stellar evolution, demonstrating how matter transforms into energy sustaining life on Earth and fueling the processes that govern stars across the universe.

Q634. Answer: B

Explanation: Saturn's spectacular ring system consists of ice and rock particles orbiting within its equatorial plane. Understanding planetary rings integrates gravity, particle dynamics, and astrophysical observation, illustrating how collisions and orbital resonance maintain structure. It highlights the diversity and beauty of planetary systems and their formation through cosmic processes observable via telescopic exploration.

Q635. Answer: C

Explanation: In 2006, the International Astronomical Union reclassified Pluto as a dwarf planet because it does not clear its orbit of other debris. Understanding this classification links astronomy, observation, and definitional criteria for planets, emphasizing scientific evolution through data refinement and the ongoing exploration of the solar system's outer regions.

Q636. Answer: A

Explanation: Venus rotates retrograde, meaning opposite most other planets' direction. Its slow, backward rotation results in long days and atmospheric dynamics unlike any other planet. Understanding

Venus's rotation integrates celestial mechanics and comparative planetology, revealing how collisions and formation events influence planetary motion and physical characteristics across solar system history.

Q637. Answer: C

Explanation: The solar corona is the outermost layer of the Sun's atmosphere, visible as a faint halo during a total eclipse. It reaches millions of degrees and emits X-rays. Understanding the corona combines astrophysics and plasma physics, explaining solar wind, magnetic fields, and their influence on Earth's space weather and technological systems.

Q638. Answer: B

Explanation: A star's color corresponds to its surface temperature—blue stars are hottest, red are coolest. Understanding stellar color links physics and spectroscopy, revealing how thermal radiation and energy output classify stars. It supports the Hertzsprung-Russell diagram, connecting temperature, luminosity, and evolution in the study of stellar properties and life cycles.

Q639. Answer: B

Explanation: Earth is located in the Milky Way Galaxy, a barred spiral system containing billions of stars, nebulae, and planetary systems. Understanding our galactic position connects astronomy, cosmology, and observation, providing context for the universe's scale and the Milky Way's structure as part of the Local Group of galaxies within the cosmos.

Q640. Answer: C

Explanation: A light-year measures distance—the span light travels in one year, about 9.46 trillion kilometers. Understanding astronomical distances connects physics and cosmology, enabling measurement of vast cosmic scales. It aids in mapping the universe, determining star locations, and analyzing the immense scale separating celestial bodies across the observable universe.

Q641. Answer: B

Explanation: Proxima Centauri, part of the Alpha Centauri system, lies about 4.24 light-years from Earth and is the closest known star beyond the Sun. Understanding nearby stars links observational astronomy and stellar evolution, providing perspective on potential exoplanetary systems and interstellar exploration prospects in future space science and technology.

Q642. Answer: B

Explanation: The Milky Way is a barred spiral galaxy with curving arms extending from a central bulge. Understanding galaxy classification integrates astrophysics and cosmology, explaining structure, star formation, and evolution processes. It helps reveal how galaxies form, interact, and change, offering insight into the dynamics shaping the universe's visible structure.

Q643. Answer: C

Explanation: The Big Bang theory proposes that the universe began from a hot, dense state about 13.8 billion years ago and has been expanding since. Understanding this theory integrates cosmology, physics, and observational evidence such as cosmic microwave background radiation, explaining universe origin, evolution, and structure through scientific observation and data interpretation.

Q644. Answer: C

Explanation: A spectroscope disperses light into a spectrum, allowing scientists to identify elements and properties of stars and galaxies. Understanding spectroscopy integrates optics and atomic physics, revealing composition, motion, and temperature. It is fundamental to modern astrophysics, enabling remote analysis of celestial bodies and their physical characteristics throughout the universe.

Q645. Answer: B

Explanation: Kepler's third law states that the square of a planet's orbital period is proportional to the cube of its distance from the Sun. Understanding this law connects mathematics, observation, and physics, providing a quantitative foundation for orbital mechanics and planetary motion within both classical and modern astronomical contexts.

Q646. Answer: B

Explanation: Parallax measures a star's apparent shift against distant background stars as Earth moves, allowing distance estimation. Understanding parallax integrates geometry and astronomy, forming the basis for stellar distance measurement. It connects observational techniques and space geometry, refining cosmic mapping and supporting calibration of broader astronomical distance scales.

Q647. Answer: A

Explanation: A neutron star forms when a massive star collapses under gravity after a supernova, compressing protons and electrons into neutrons. Understanding neutron stars combines astrophysics and nuclear physics, revealing extreme states of matter, density, and magnetism that expand knowledge of stellar evolution and the physics governing compact celestial objects.

Q648. Answer: B

Explanation: The redshift of distant galaxies indicates they are moving away, implying universal expansion. Edwin Hubble's observations confirmed this relationship. Understanding redshift integrates light physics and cosmology, providing empirical support for the Big Bang theory and revealing the dynamic nature of space-time shaping the cosmos across billions of years.

Q649. Answer: B

Explanation: Venus rotates very slowly—one rotation takes 243 Earth days, while its orbit around the Sun takes only 225 days. Understanding this phenomenon integrates planetary dynamics and rotational physics, highlighting the variety of motion and atmospheric conditions across planets and providing insight into early solar system evolution and planetary interactions.

Q650. Answer: C

Explanation: Current cosmological measurements, based on the cosmic microwave background and expansion rate, estimate the universe's age at about 13.8 billion years. Understanding cosmic age integrates physics, astronomy, and mathematics, revealing the universe's evolution from the Big Bang to present structure through precise observational and theoretical astrophysical analysis.

Q651. Answer: B

Explanation: The Sun is composed primarily of hydrogen, which undergoes nuclear fusion to form helium, releasing vast amounts of energy. Understanding solar composition integrates astrophysics and nuclear chemistry, explaining how hydrogen fusion sustains the Sun's luminosity and influences stellar life cycles. It provides insight into stellar evolution and the processes powering stars across the observable universe.

Q652. Answer: C

Explanation: Mars appears red because its surface is rich in iron oxide, or rust. This composition and thin atmosphere distinguish it from other terrestrial planets. Understanding Mars' geology integrates planetary science, mineralogy, and atmospheric physics, offering insights into planetary evolution, water history, and the potential for past or future habitability within our solar system.

Q653. Answer: B

Explanation: The Andromeda Galaxy (M31) is a spiral galaxy similar to the Milky Way, featuring a central bulge and winding arms of stars and dust. Understanding Andromeda helps astronomers study galactic structure, rotation, and evolution. It connects astrophysics and cosmology, offering insight into galaxy formation, mergers, and our galaxy's eventual interaction with Andromeda in the distant future.

Q654. Answer: C

Explanation: Jupiter's immense size and rapidly rotating metallic hydrogen core generate the strongest magnetic field among planets. Understanding planetary magnetism integrates geophysics and plasma physics, revealing interactions with solar wind and auroral phenomena. Jupiter's magnetic field provides a model for studying magnetospheres, planetary interiors, and magnetic processes across the solar system and beyond.

Q655. Answer: B

Explanation: Elliptical galaxies are composed mainly of older stars with minimal gas and dust, resulting in low star formation. Understanding galactic morphology integrates astrophysics and cosmology, showing how mergers and evolution affect star populations. It illustrates how galaxy structure and composition reflect the universe's history and the aging process of stellar systems over billions of years.

Q656. Answer: B

Explanation: The average Earth–Sun distance is about 93 million miles, or 1 astronomical unit (AU). Understanding this distance is fundamental to celestial mechanics and scale models of the solar system. It provides a standard unit for measuring planetary orbits and comparing distances, reinforcing the geometric and observational foundations of planetary and space science.

Q657. Answer: C

Explanation: The Sun is a G2V main-sequence star, generating energy by hydrogen fusion in its core. Understanding stellar classification integrates astrophysics and thermodynamics, revealing how temperature, luminosity, and composition determine a star's spectral type. The Sun serves as a model for studying stellar evolution, energy generation, and the life cycle of similar medium-sized stars.

Q658. Answer: B

Explanation: A planet's orbit is its curved path around the Sun due to gravitational attraction. Understanding orbital motion connects celestial mechanics, gravity, and motion laws. It provides the basis for predicting planetary positions, explaining retrograde motion, and calculating spacecraft trajectories, fundamental to astronomy, astrophysics, and space exploration dynamics across planetary systems.

Q659. Answer: C

Explanation: Jupiter's atmosphere is composed primarily of hydrogen, with helium and traces of methane and ammonia. Understanding gas giant composition integrates planetary science and spectroscopy, explaining how light and mass distribution reveal atmospheric structure. It highlights solar system formation processes and the distinction between gas giants and terrestrial planets in chemical makeup.

Q660. Answer: C

Explanation: Halley's Comet is a periodic comet with a 76-year orbit around the Sun. It is composed of ice, dust, and rock that vaporize near the Sun, forming tails. Understanding comets integrates astronomy and chemistry, providing insight into early solar system materials and illustrating how cometary orbits reveal gravitational influences and solar interactions.

Q661. Answer: B

Explanation: Retrograde motion occurs when Earth overtakes outer planets in orbit, making them appear to move backward temporarily. Understanding this optical illusion integrates celestial geometry and observation, demonstrating relative motion effects and confirming the heliocentric model. It enhances comprehension of planetary dynamics and visual phenomena within solar system observation and modeling.

Q662. Answer: A

Explanation: Mars has a rotation period of about 24.6 hours, closely matching Earth's day length. Understanding planetary rotation integrates comparative planetology and geophysics, showing how angular momentum and formation history influence rotation rates. This similarity helps researchers study climate cycles, seasonal patterns, and surface conditions relevant to habitability and exploration.

Q663. Answer: B

Explanation: The photosphere is the Sun's visible surface layer, where light escapes into space. Understanding the photosphere integrates radiation physics and observational astronomy, explaining sunspots, convection, and solar brightness. It highlights the Sun's structure and energy transport, forming the foundation for studying solar activity and its effects on space weather.

Q664. Answer: C

Explanation: Uranus rotates on its side, likely due to a massive early collision, resulting in extreme seasonal variations. Understanding Uranus's tilt integrates celestial mechanics, planetary evolution, and dynamics. It provides insight into solar system formation, axial stability, and the diversity of planetary orientations influencing climates and orbital characteristics across planets.

Q665. Answer: B

Explanation: Reflecting telescopes use mirrors to gather and focus light, reducing distortion and allowing larger apertures than refractors. Understanding telescope design integrates optics and engineering, illustrating how observation technologies enhance resolution and sensitivity. Reflectors revolutionized astronomy, enabling deep-space study of galaxies, nebulae, and cosmic phenomena across electromagnetic wavelengths.

Q666. Answer: B

Explanation: Absolute magnitude measures a star's intrinsic brightness as if viewed from 10 parsecs away. Understanding luminosity metrics integrates astrophysics and photometry, distinguishing observed brightness from distance effects. It aids in determining stellar size, energy output, and evolution, allowing astronomers to compare stars and construct models of stellar populations and galactic structures.

Q667. Answer: B

Explanation: A supernova is the cataclysmic explosion marking the death of a massive star, releasing immense energy and heavy elements into space. Understanding supernovae connects nuclear physics, cosmology, and stellar evolution, explaining element formation and galactic enrichment processes. These explosions seed future star systems and contribute to the universe's chemical diversity and evolution.

Q668. Answer: B

Explanation: Mercury experiences extreme temperature swings due to its lack of atmosphere and slow rotation, ranging from scorching days to freezing nights. Understanding thermal variation connects planetary physics, energy transfer, and atmospheric science, illustrating how atmosphere and rotation affect climate and habitability conditions across solar system bodies.

Q669. Answer: A

Explanation: The Doppler effect causes shifts in light frequency from moving objects—redshift indicates recession, blueshift indicates approach. Understanding this principle integrates wave physics and cosmology, providing evidence for universe expansion and star motion. It is essential for measuring galaxy velocities, binary star systems, and exoplanet detection across astrophysical observations.

Q670. Answer: C

Explanation: The event horizon marks the limit of escape velocity equal to light speed, trapping all matter and radiation. Understanding black holes combines relativity and astrophysics, revealing spacetime curvature and gravity's extremes. It informs theoretical models and observational evidence, deepening comprehension of cosmic phenomena and the universe's fundamental physical laws.

Q671. Answer: B

Explanation: The Andromeda Galaxy (M31) is approaching the Milky Way and is expected to merge in about 4 billion years. Understanding galactic collisions integrates cosmology and gravitational dynamics, explaining structure formation and evolution. It highlights how interactions reshape galaxies and star systems, contributing to the dynamic and ever-changing nature of the universe.

Q672. Answer: A

Explanation: Venus's dense atmosphere, rich in carbon dioxide and sulfuric acid clouds, creates an intense greenhouse effect, making it the hottest planet. Understanding Venus's atmosphere integrates chemistry, physics, and climatology, illustrating greenhouse dynamics and their planetary consequences relevant to comparative climate and Earth's atmospheric science.

Q673. Answer: B

Explanation: The solar nebula was a cloud of gas and dust that collapsed under gravity, forming the Sun and planets through accretion. Understanding nebular theory integrates physics, chemistry, and planetary science, explaining solar system formation and composition. It connects cosmic processes of star and planet creation observed throughout galaxies.

Q674. Answer: D

Explanation: A star's color reveals its surface temperature—blue stars are hottest, red stars coolest. Understanding this relationship integrates thermodynamics and spectroscopy, allowing classification using the Hertzsprung-Russell diagram. It forms the basis for studying stellar evolution, composition, and the energy output governing stellar behavior throughout their life cycles.

Q675. Answer: B

Explanation: The cosmic microwave background (CMB) is the residual radiation from the universe's early hot phase, providing strong evidence for the Big Bang. Understanding CMB integrates cosmology and particle physics, revealing early universe conditions, density fluctuations, and structure formation. It underpins modern cosmological models explaining the universe's origin, evolution, and large-scale geometry.

Q676. Answer: B

Explanation: The scientific method is a flexible framework involving observation, hypothesis formulation, experimentation, and analysis. It encourages iterative learning and revision based on evidence. Understanding this process demonstrates how scientists build reliable knowledge, emphasizing inquiry, objectivity, and the systematic pursuit of truth across disciplines through continuous evaluation and peer verification.

Q677. Answer: B

Explanation: Isaac Newton developed the three laws of motion and the law of universal gravitation, forming the foundation of classical physics. His work unified celestial and terrestrial mechanics under one framework. Understanding Newton's contributions highlights the progression of scientific thought and the application of mathematics to explain natural phenomena accurately and predictively.

Q678. Answer: C

Explanation: Thomas Kuhn introduced the concept of paradigm shifts, arguing that science advances through revolutionary changes rather than gradual accumulation. His work emphasized the social and conceptual context of science. Understanding Kuhn's theory underscores how transformative discoveries redefine frameworks and encourage critical reflection on the nature and progress of scientific inquiry.

Q679. Answer: B

Explanation: Reliability depends on consistent measurement techniques and proper instrument calibration. This ensures that repeated measurements yield stable, dependable results. Understanding reliability connects to the broader principles of assessment validity, precision, and reproducibility—core features distinguishing scientific evidence from opinion or bias within experimental and educational measurement contexts.

Q680. Answer: B

Explanation: Falsifiability, introduced by Karl Popper, defines a scientific claim as one that can be tested and potentially disproven. This criterion separates science from pseudoscience. Understanding falsifiability reinforces the critical evaluation of evidence, ensuring that hypotheses remain testable, transparent, and open to revision under new empirical findings.

Q681. Answer: C

Explanation: Ethical guidelines ensure honesty, transparency, and respect for research participants. They prevent harm, falsification, and exploitation while promoting accountability. Understanding ethics in science connects moral philosophy and professional practice, emphasizing that the pursuit of knowledge must balance innovation with responsibility, fairness, and societal welfare.

Q682. Answer: B

Explanation: Peer review ensures that scientific research is evaluated for accuracy, validity, and significance before publication. Experts assess methodology, results, and conclusions. Understanding this process highlights collaborative verification as essential to maintaining scientific integrity, fostering credibility, and ensuring that knowledge builds on rigorously tested and trustworthy foundations.

Q683. Answer: A

Explanation: Galileo pioneered experimental methods and precise observation, transforming science from speculation into empirical investigation. His use of instruments and mathematical reasoning revolutionized physics. Understanding Galileo's contribution underscores the shift toward evidence-based inquiry, forming a cornerstone of the modern scientific method and data-driven experimentation.

Q684. Answer: B

Explanation: A scientific theory explains phenomena through extensive evidence and repeated testing, while a hypothesis is a specific, testable prediction. Understanding this distinction reinforces how science organizes knowledge hierarchically, showing how theories evolve from empirical success and serve as frameworks guiding ongoing investigation and discovery.

Q685. Answer: B

Explanation: Reproducibility ensures that results can be independently confirmed, strengthening trust in scientific conclusions. It demonstrates that findings are not due to chance or bias. Understanding reproducibility highlights the collaborative nature of science and its dependence on open methods, verifiable data, and consistent standards for evaluating truth claims.

Q686. Answer: B

Explanation: Empirical evidence refers to information gathered through observation, experimentation, and sensory experience. Understanding this concept highlights the foundation of the scientific method, emphasizing that reliable knowledge must be observable and measurable. It distinguishes science from belief-based systems and underscores objectivity and verifiability in research practices.

Q687. Answer: B

Explanation: Einstein's relativity redefined fundamental concepts of space, time, and gravity, replacing Newton's framework under extreme conditions. This transition represents a paradigm shift, showing how scientific revolutions occur when new theories better explain anomalies. Understanding this shift reveals the evolving, self-correcting nature of science and its pursuit of deeper explanatory power.

Q688. Answer: A

Explanation: Validity indicates how accurately a measurement reflects the concept it is intended to assess. Understanding validity is central in both scientific and educational evaluation. It ensures meaningful interpretation of results, linking assessment design, operational definitions, and theoretical constructs to the accurate measurement of real-world phenomena.

Q689. Answer: B

Explanation: Scientific models simplify and represent complex systems to enhance understanding and prediction. They serve as conceptual or mathematical tools linking theory and observation. Understanding modeling underscores abstraction and simulation as integral to scientific reasoning, fostering insight into processes that are too vast, small, or intricate for direct study.

Q690. Answer: B

Explanation: Statistics help scientists interpret data objectively, distinguishing meaningful patterns from random variation. Understanding statistical reasoning integrates mathematics and empirical analysis, reinforcing the importance of probability, inference, and replication in validating results. It ensures that conclusions reflect genuine effects rather than chance, strengthening scientific rigor and transparency.

Q691. Answer: C

Explanation: Occam's Razor advises choosing the simplest explanation that fits all observations, avoiding unnecessary assumptions. Understanding this principle encourages logical economy and clarity in hypothesis formation, fostering efficient scientific reasoning. It highlights how simplicity and precision aid in developing models that are both predictive and conceptually elegant.

Q692. Answer: B

Explanation: Scientific skepticism encourages questioning evidence and methods to ensure conclusions are justified and replicable. Understanding skepticism emphasizes the importance of critical thinking and methodological rigor. It helps maintain objectivity, preventing acceptance of false claims and supporting science's self-correcting nature through continuous scrutiny and refinement.

Q693. Answer: B

Explanation: The scientific revolution emphasized empirical methods, experimentation, and mathematical reasoning, transforming natural philosophy into modern science. Understanding this historical period connects to key figures like Galileo, Newton, and Bacon, who shaped scientific inquiry into a systematic, evidence-driven process forming the foundation of contemporary science and technology.

Q694. Answer: B

Explanation: Double-blind experiments conceal treatment conditions from both participants and researchers, minimizing bias in data collection and interpretation. Understanding this method illustrates the importance of control and impartiality in experimental design, ensuring that results reflect true effects rather than expectations or preconceived notions.

Q695. Answer: B

Explanation: Calibrated instruments and standardized procedures minimize measurement error, ensuring data accuracy and comparability. Understanding error reduction integrates physical measurement principles and assessment reliability. It demonstrates how consistency, precision, and methodological discipline underpin trustworthy data collection, enhancing the overall validity and credibility of scientific findings.

Q696. Answer: D

Explanation: Peer review and replication validate research findings through independent verification and critical analysis. Understanding credibility highlights the community-based nature of science, where transparency, repeatability, and scrutiny safeguard against error and bias. This process ensures knowledge reliability and promotes ongoing refinement of theories and methods in scientific practice.

Q697. Answer: B

Explanation: Quantitative research uses numerical data to identify patterns, while qualitative research explores meanings and experiences through description. Understanding their difference emphasizes methodological diversity in science and social inquiry, showing how each approach enriches understanding, supports triangulation, and fosters comprehensive explanations of complex natural and human phenomena.

Q698. Answer: B

Explanation: The replication crisis highlights challenges in reproducing prior research due to methodological flaws or publication bias. Understanding replication underscores transparency, data sharing, and methodological rigor as pillars of credible science. It has prompted reforms in research design, emphasizing open science practices and stronger standards for evidence validation across disciplines.

Q699. Answer: B

Explanation: Social accountability means scientists must consider the ethical, environmental, and societal implications of their work. Understanding this concept links science to human values and policy, promoting responsible innovation. It encourages reflection on how discoveries affect public welfare, sustainability, and the broader consequences of technological and scientific progress.

Q700. Answer: D

Explanation: Open access publishing removes financial barriers, allowing broader dissemination of scientific knowledge. Understanding open access connects to equity, collaboration, and the democratization of information, enhancing global participation in research. It supports transparency, accelerates innovation, and reinforces science as a shared, cumulative endeavor benefiting society at large.

Made in the USA
Monee, IL
07 November 2026

68054640R00129